Engineers Ireland

CALLED TO SERVE

Presidents of the Institution of Civil Engineers of Ireland 1835-1968

Ronald Cox & Dermot O'Dwyer

Published 2014 by:

Engineers Ireland

22 Clyde Road

Dublin 4

Copyright:

Engineers Ireland, 2014

All rights reserved. No part of this publication may be reproduced, stored in a retrieval system, or transmitted in any form or by any means, electronic, mechanical, photocopying, recording or otherwise, without the prior permission of the publishers.

ISBN: 978-09502874-1-6

This book has been sponsored by the ESB

Printed by Read's, Sandyford, Dublin 18

Contents

Introduction

1   The Institution

2   The Presidential Addresses

3   Biographical Sketches of Presidents

Reference Sources

Authors

Ronald Cox is a Chartered Engineer and a Research Associate in the Department of Civil, Structural & Environmental Engineering at Trinity College Dublin. He was formerly a Senior Lecturer in Civil Engineering and one-time Dean of Engineering at Trinity College Dublin.

Dr Cox is a Member of the Institution of Civil Engineers, a Fellow of the Institution of Engineers of Ireland, a Fellow of the Irish Academy of Engineering, a Member of the American Society of Civil Engineers, and the current chairman of the Engineers Ireland Heritage Society.

Recent publications include *Civil Engineering Heritage: Ireland* (1998), *Ireland's Bridges* (2003), *Engineering Ireland* (2006), and *Ireland's Civil Engineering Heritage* (2013).

Dermot O'Dwyer is a Chartered Engineer and an Associate Professor in the Department of Civil, Structural & Environmental Engineering at Trinity College Dublin. He is a Member of the Institution of Civil Engineers and of the Institution of Engineers of Ireland. Professor O'Dwyer represents Engineers Ireland on the ICE Panel for Historical Engineering Works and is a member of the Editorial Panel for the peer-reviewed journal *Engineering History and Heritage.*

'Reading makes a full man, writing an exact man, and speaking a ready man'
(Francis Bacon)

Introduction

Past Presidents of the Institution of Civil Engineers of Ireland (ICEI) have interpreted their role in a variety of ways, including chairing meetings of the council, presiding at the presentation of technical papers, and representing the Institution and the engineering profession, both nationally and internationally. The tradition of presenting a presidential address during the term of office of each president was begun in 1856 by George Willoughby Hemans. Since that time, with few exceptions, each president has addressed the membership in their own individual style, often drawing on the experiences gained during their respective careers.

The feelings of the presidents faced with this tradition can best be illustrated by quoting Alfred Delap (President 1927) on the subject:

> 'The whole question of a Presidential address is a very difficult one, and one on which very little, if any, help can ever be hoped for. Past Presidents are certainly not going to help their successors to evade or avoid an ordeal which *they* have successfully gone through; future or potential Presidents naturally feel that such a simple matter will give them no trouble when their time comes, and so your President is left in isolation to face the inevitable'.
>
> 'But is a Presidential Address inevitable?'
>
> 'Our Institution was founded in 1835, and the first Presidential address was, I believe, not delivered till 1856, by Hemans, and from time to time since, the value of the Presidential addresses has been seriously questioned. (The speaker will feel that he has not lived, or spoken, in vain if his address finally convinces the Institution that it would be wiser to let its Presidents preside in as nearly absolute silence as possible)'.
>
> 'It must be faced that an address from your President is customary and, therefore, necessary, as he has not sufficient courage to open his year of office by ignoring, or departing from, the recognised custom. (Besides, a time will, I hope, come when I shall have the opportunity of seeing others in the embarrassing position, in which I now stand, of having to address my fellow Members on some subject, while feeling very strongly that I know nothing about anything that can possibly interest anyone)'.

In this book, the presidential addresses have been analysed in an effort to present their highlights and insights in an historical context. The full presidential addresses were published in the Transactions of the Institution, which are now available online in digital format at http://digitalcollections.tcd.ie/home/. It is hoped that the analyses presented in this publication will serve to whet the appetite for a more detailed study of the individual addresses.

The analyses of the presidential addresses are preceded by a brief history of the ICEI up to its unification with the Engineers Association (Cumann na nInnealtóirí) in 1969. This is presented in order to provide a framework within which the presidential addresses may be considered. It consists of extracts from the fuller treatment of the history published previously in Cox (2006).

A major portion of the book is devoted to short biographies of each of the presidents - from John Fox Burgoyne in 1835 to James Clement Ignatius Dooge in 1968.

The support for the project of the members of the Heritage Group of the Irish Academy of Engineering is gratefully acknowledged.

Chapter One
The Institution

On Thursday 6 August 1835, a meeting of twenty civil engineers was held at the office of the Board of Public Works (Ireland), located at the time in the Custom House, Dublin, and presided over by Colonel (later Field Marshal Sir) John Fox Burgoyne, chairman of the Board. It appears that the meeting was a sequel to a preliminary meeting at which a further sixteen engineers were in attendance and signified their support for the formation of a society for their own improvement. The number of founder members thus totalled thirty-six, composed mainly of engineers working for the Board.

Burgoyne pointed out that the profession of civil engineering had been at a low ebb in Ireland and that persons without education or skill had frequently been employed in operations of importance, resulting in bad or injudicious works, wasteful or fruitless expenditure and a certain degree of discredit to the country. He explained that they were now adopting the measure best calculated to prevent the recurrence of these evils, by organising a society, which it was hoped might be the means of adding respectability to the profession of civil engineers in Ireland and rendering some service to the country. Thus came into being *The Civil Engineers Society of Ireland* - a name, however, that was soon to be changed. The society had for its object 'the promotion of science in general, but more particularly as connected with the profession of Civil Engineers'.

A general meeting of the society was held on 17 August 1844 in the Custom House with Burgoyne as chairman and Robert Mallet acting as secretary, when it was resolved that 'The Institution of Civil Engineers of Ireland be formed for the promotion of mechanical science and more particularly for the acquisition of that species of knowledge which constitutes the profession of a Civil Engineer'. **The Institution of Civil Engineers of Ireland** (ICEI) was to retain this title for the next 125 years. Burgoyne remained as president and three vice-presidents and ten other members constituted the council. Most of the credit for the rejuvenation of the 'Institution' was due to Mallet and it was he who prepared a Code of By-Laws that placed the ICEI on a firm footing.

In those early years, the Board of Public Works provided an office and meeting rooms in the Custom House, although it would seem that on occasions, meetings were held in the rooms of the Geological Society. Regular meetings were held until the Board had to terminate the arrangement when the rooms were required for the accommodation of a large emergency staff recruited to administer relief measures following the Great Famine. Thus, in 1853, the headquarters of the ICEI were moved to apartments rented from the Richmond Institute for the Blind at 41 Upper Sackville Street (as O'Connell Street in Dublin was then named). This change proved a strain on the resources of the organisation, partly due to the cost of furnishing the new premises and maintenance charges out of all proportion to the size of the membership. Membership had been falling owing to a change in government policy at the time and the lessening of the provision for public works, with consequential lack of work for engineers.

Although many administrative and structural problems had arisen, the essential aims and objects of the founders were being fulfilled. Meetings were held frequently and papers presented and discussed. A good deal of time was devoted to the description of models and to the demonstration of engineering instruments, as the era of technical journals and commercial shows and exhibitions was in its infancy. Through the generosity of individual members, there was already a considerable accumulation of books and models for which rooms had to be provided and a curator appointed. Sadly, the models were later dispersed and only remnants of the library collection have survived to form an archive, now housed at 45 Merrion Square under an arrangement with the Irish Architectural Archive. The Transactions (technical papers) of the ICEI were first published at the end of the session 1844-45. They contain in individual papers, and the often substantial presidential addresses, an impressive record of the contribution that engineers have made to the economy, to the welfare of the community, and to the country as a whole.

The perilous seven years with headquarters at Sackville Street were ended by a most magnanimous gesture by the Board of Trinity College Dublin who undertook, at the request of the Professor of Civil Engineering, Samuel Downing, to house and safeguard the collection of models and library items and to provide suitable rooms for council and general meetings, including fuel and light, free of charge and without limitation as to period. Trinity College was to be used for the meetings of the ICEI for the next thirty years, whilst the Council considered ways of providing for an office and library in a business part of town. In due course, apartments were rented in the upper part of 136 St. Stephen's Green West, which was accomplished by a combination of increasing members' subscriptions and a bequest of £1,200 (together with a collection of professional books), received under the will of a Past-President, Michael Bernard Mullins. The five years, from January 1874 to November 1879, when the Reading Room and Offices were located at St. Stephen's Green, marked a most significant phase in the history of the ICEI, for it was during this period that it was granted a Royal Charter of Incorporation, thus giving it real status as a body entitled to represent and act for the engineering profession in Ireland.

The necessity of obtaining a Charter had been felt since the inception of the ICEI, and the granting of a Charter in 1877 marked a new phase in its growth. The then President, Robert Manning was named in the Charter together with others who 'have formed themselves into a Society for promoting the acquisition of that species of knowledge which appertains to the professions of Civil and Mechanical Engineers, and for the advancement of Engineering and Mechanical Science'. It will be observed that the original Charter stated that the objects were the advancement of Engineering Science as well as Mechanical Science, which latter was alone mentioned in the 1844 constitution, whilst the profession of Mechanical Engineer is mentioned for the first time. This may be accounted for by the fact that a number of prominent mechanical engineers were members of the Council at that time, but in any case it would seem as if it were intended that two distinct professions would be catered for by the one organisation. The retention of the word 'Civil' in the name appears to reflect the meaning of the word, which encompassed at that time mechanical sciences, rather than the exclusion of what is now a sister profession. The wisdom and foresight of those who sought and obtained the Charter ensured the independence of the profession and control by its own members.

Within two years of the granting of the Charter, the ICEI once again found itself in difficulties with accommodation. The lease of the house at St. Stephen's Green had expired and the landlord's requirements for renewal were unacceptable. Ultimately, a short unexpired term of the interest of the lease of 35 Dawson Street was purchased. Some of the rooms were already let to tenants and there was a garden with stables at the rear that was to eventually become the site for a new Lecture Hall and Supper Room. The Office and Library was transferred to Dawson Street at the end of 1879, but the general meetings of the ICEI were still held in Trinity College.

The members felt that the progress of the ICEI was being retarded and its usefulness greatly lessened by the general meetings not being held in a hall connected with the headquarters of the organisation. Eventually, a hall was built in the rear garden of the Dawson Street premises and the first general meeting to be held there was on the evening of 16 December 1891, when the grateful thanks of the ICEI was again tendered to the Provost and Fellows of Trinity College for their hospitality in affording the organisation facilities for holding its meetings within the college for such a long period. The Supper Room was added in 1899. Meetings were held here in the Hall on the first Monday of the month from November to May when learned papers on a variety of engineering topics were delivered and discussed, that is, except for the Presidential Address, a custom first introduced in 1856, from which, according to precedent and practice, discussion is precluded.

The fundamental principles set out in the original Charter stood the test of time and continued to be valid and applicable under the Constitutions of 1922 and 1937 by virtue of the Adaptation of Charters Act 1926 that provided, inter alia, that any board or body governed by Charter shall be deemed a board or body constituted by statute.

By the mid-twentieth century the profession of engineering had so advanced and the scope of engineering science had become so enlarged that the definition in the original Charter required extension. Furthermore, the management of the affairs of the ICEI demanded a larger and more representative Council. The Institution of Civil Engineers of Ireland (Charter Amendment) Act 1960 provided for a Council, exclusive of Officers, of not less than 21 persons nor more than the number prescribed in the By-Laws. The setting up of Committees was now greatly facilitated with the increase of personnel and the presence of a young and vigorous membership. The purpose of the ICEI was redefined as 'for the promotion of that species of knowledge which appertains to the profession of Engineering and the special advancement of Engineering Science'. The word 'Civil' was to remain in the Institution's title for another decade, although its application extended far beyond the narrow confines of civil engineering.

In the 1960s the most significant development, and the most dramatic in its consequences, was the setting up of a Policy Committee to consider the ways and means whereby the ICEI might serve the interests of members of all branches of engineering in Ireland. The committee investigated the benefits that engineers derived from the wide range of societies to which they contributed. Representatives of other engineering societies were consulted about the possibility of creating more dynamic engineering opinion and development in the country. Various study groups were set up to examine the best means of achieving unified representation for the engineering profession in Ireland on matters concerning professional standards and international mutual recognition under the Treaty of Rome and the EEC (now the EU) The effect of the newly created Council of Engineering Institutions (CEI) in the United Kingdom had also to be taken into account.

A special General Meeting of the Institution was held on 20 April 1964 at which a report was presented entitled *'Institution: A Unified Society Plan for Development 1964'*. The report was adopted in principle and the Council was authorized to purchase a site for the construction of an Engineering Centre and the disposal of the premises at 35 Dawson Street.

The disposal of the premises in Dawson Street was intended to lead to the building of or the acquisition of new modern premises and the setting up of an adequate Engineering Centre that, it was hoped, would be shared with Cumann na nInnealtóirí (CnaI), often referred to as The Engineers' Association. The centre was to provide a well-equipped modern lecture theatre as well as offices. The search for a building or a site proved to be a very long and fruitless task. The government vetoed what was considered to be an ideal site, whilst other sites failed to be granted planning permission on appeal.

In order to cater for the needs of all engineering disciplines in the country, it was stated that the intention of the Institution was to seek an Act to change its name to 'The Institution of Engineers of Ireland' and it was explained that, whilst up to that time, it had been open to all branches of the profession, it had catered mainly for the requirements of the civil engineering branch. So, it was felt necessary to seek the change of name so as to draw in a more widely-based membership. The organization was to become more deeply involved in assisting, through the development of technology, in the nation's economic and industrial growth by reorganizing the structure of the engineering profession so as to embrace all its branches.

The two distinct functions of the ICEI as a 'learned society', and the Cumann catering for the welfare of engineers collectively and individually, were clearly enunciated, the hope being expressed that one day the two bodies might draw closer together. The early history of the 'Cumann' from its formation in 1928 has been described by Finbar Callanan in Cox (2006).

The reason for an organisation separate from the ICEI was that the council of the ICEI considered that its charter precluded it from negotiating on conditions of employment. Their work was crowned by the unification of the Irish engineering profession in 1969, when the administration and structure of the Cumann, which had developed so well in the previous decades, provided much of the bedrock administration and structure of the unified Institution of Engineers of Ireland (IEI).

The lack of a suitable organ of communication for the members of the Cumann had been a long-felt disadvantage and in December 1940 the first issue of the *Engineers Journal* appeared - initially as an annual, later as a quarterly, and subsequently in 1949 as a monthly publication. *"The Journal"* as it was affectionately known, provided innumerable articles on a comprehensive range of subjects pertaining to engineering history, education, practice and management as well as a very wide range of engineering news, views, comment and special features.

In 1939, a Memorandum, Articles of Association and By-laws were drafted and the aims of the Cumann were redefined and expanded. This new Constitution became effective in May 1941 and was to govern the Cumann to the end of its existence. In 1941, a Trade Union Bill was passed by Dail Éireann making it illegal for any body, not being an exempted body, to carry out negotiations for the fixing of wages or conditions of employment unless such body was the holder of a negotiating licence. The Cumann became an exempted body by ministerial order and thus became empowered to negotiate on behalf of its members. In 1955, the Cumann moved to headquarters premises at 22 Clyde Road in Dublin 4.

At that stage of the growth and development of the Cumann, considerable progress had been made in fulfilling its objective of advancing the standing, status and remuneration of members of the profession. This work was pursued actively by a system of vocational groups, which represented the various areas of employment, and which had developed over the years. Notable amongst the groups in question were those representing engineering personnel in the ESB, the Local Authorities, the Defence Forces (engineer officers only), Bord na Mona, Posts & Telegraphs and CIE. Additionally the Cumann played a very positive role in furthering the interests of smaller groups and individuals for whom negotiating services were supplied. Remuneration was consistently improved and without question the efforts of the Cumann in its formative years and subsequently, played a very significant part in advancing the cause of engineering in Ireland and in projecting the profession as a rewarding career to be sought and followed.

The Regions were particularly valuable sounding boards for the members throughout the country, where ideas were generated, proposals were made, and grievances and queries aired. It was a significant strength of the Cumann that it was in constant communication with its members and its ability to deal with bread-and-butter issues as well as furthering a wide range of policy issues and educational and social events was a unifying influence that added greatly to its growth throughout the 1950s and 1960s in particular.

When the Cumann was set up in 1928 many engineers felt that the ICEI should have undertaken the tasks for which the Cumann was established, by modifying its rules or by seeking to have its Charter amended. This feeling persisted down through the years and the relationship between both bodies was not always equitable. About two-thirds of ICEI members were also members of the Cumann, many serving as officers or council members of both organizations. It was inevitable, as the Cumann grew in numbers and influence, that the apparent encroachment by the Cumann on what would have been considered ICEI territory, became a cause of friction. There was also the ongoing problem of two separate organisations speaking on behalf of the profession, each with its own point of view and not necessarily in agreement with each other.

However, both bodies had a common purpose in seeking registration, which for decades had been a fundamental objective of both organisations, and in 1965, this matter was again raised with the government. In that year also the ICEI decided to sell its premises in Dawson Street and their offices were moved temporarily to the Intercontinental Hotel in Ballsbridge. Once again the academic world came to the rescue, University College Dublin permitting meetings to be held, free of charge, in the Engineering School in the Science Buildings at Upper Merrion Street (now the Department of the Taoiseach). The Cumann agreed to provide space for the ICEI's library at their headquarters at 22 Clyde Road in Ballsbridge.

In essence, the coming together of the two organisations that had been simmering for some years, began to appear more rational than it had ever been before. The following motions were placed before the central council of the Cumann, and passed unanimously: 'that this meeting is of the opinion that unity is desirable between Cumann na nInnealtóirí and The Institution of Civil Engineers of Ireland and that, as a step in that direction, arrangements should be made for the establishment of a joint secretariat in the existing circumstances; and while both bodies would continue to act independently, the councils of both bodies should explore jointly the possibility of the unification of both societies'. Within a few weeks, the council of ICEI had passed identical motions.

Progress on unification proceeded apace and a joint executive of the Cumann and the ICEI was formed, under the chairmanship of the chairman of the central council, Finbar Callanan, and spokesman for the Cumann, to coordinate the task of unification and to keep the members of both organisations fully informed by way of meetings in every Region and detailed reporting in the Journal. A proposed organisational structure for the unified Institution had already been developed by the Unification Committee, which was representative of both bodies and was led by the dynamic former Cumann President, Jock Harbison. It was agreed that the best basis for unification would be a modification of the Charter of the Institution of Civil Engineers of Ireland (which required legislation) and the transfer to the new body of the Cumann's negotiating authority. This meant in fact that the ICEI (albeit under a change of name) would continue in existence and that the Cumann would cease its separate existence.

The best legal advice was obtained and the engineering senators were most helpful, especially Senator James Dooge, whose guidance and constant advice and assistance was invaluable to the successful outcome of the process within the Oireachtas of securing the Charter Amendment Act which governed the setting up of the unified body. The first joint meeting of the two councils in 1968 was chaired jointly by Patrick Raftery, president of the ICEI, and Finbar Callanan, chairman of the joint executive of the Cumann and ICEI. There was no dissent whatever, and from then on progress towards unification seemed inevitable, and so it turned out to be, the necessary amending Act being passed in May 1969.

The passing of the Institution of Civil Engineers of Ireland (Charter Amendment) Act 1969 provided for the formation of a new body to represent the engineering profession in Ireland. As well as widening the range of activities, the Act embraced most areas of specialisation in engineering and provided an umbrella and platform for the exploitation and development of these various specialisms, and, most importantly, combined the aims and objectives of both the Cumann and the ICEI.

Having celebrated its 40th anniversary in 1968, the Cumann, on 12 May 1969, with the full agreement of its members to the new charter, by-laws, structure, governance and title of the new organisation, ceased as such to exist. The Cumann did, however, survive in the memory of all who involved themselves in its continual growth during the four previous decades and in the almost 2,000 members of all disciplines who brought such vigour to the new body. It also survived in the title of the new professional body, which in the Charter Amendment Act states that the title of the body shall be **The Institution of Engineers of Ireland (IEI)** and in the Irish language **Cumann na nInnealtóirí**. The new professional body was recognized by Act of the Oireachtas as the sole body licensed to award the title 'Chartered Engineer' within the State, and to maintain a register of Chartered Engineers practising in Ireland.

ICEI Presidents 1835 – 1968

1835	John Fox Burgoyne		1921	Joshua Hargrave
1845	John Radcliff		1922	Pierce Francis Purcell
1846	Harry David Jones		1924	James Thomas Jackson
1850	Richard John Griffith		1926	Arthur Hassard
1856	George Willoughby Hemans		1927	Alfred Dover Delap
1859	Michael Bernard Mullins		1929	Michael James Buckley
1861	Richard John Griffith		1930	Joseph Mallagh
1863	Charles Blacker Vignoles		1931	Stephen Gerald Gallagher
1865	Robert Mallet		1932	Laurence Joseph Kettle
1867	William Anderson		1934	Nicholas O'Dwyer
1869	John Ball Greene		1936	Frank Sharman Rishworth
1871	Bindon Blood Stoney		1938	Joseph Albert Ryan
1873	Charles Philip Cotton		1940	Henry Nicholas Walsh
1875	Alexander McDonnell		1942	Thomas Joseph Monaghan
1877	Robert Manning		1943	Thaddeus Cornelius Courtney
1879	John Bailey		1944	Patrick Joseph Raftery
1881	Parke Neville		1945	Norman Albert Chance
1883	William Hemingway Mills		1946	John Purser
1885	John Audley Frederick Aspinall		1947	Joseph MacDonald
1887	John Purser Griffith		1948	Joseph Phelan Candy
1889	Spencer Harty		1949	Michael Anthony Hogan
1891	Thomas Francis Pigot		1950	Thomas Aloysius McLaughlin
1893	John Chaloner Smith		1951	William Ian Sidney Bloomer
1895	James Price		1952	Patrick George Murphy
1896	James Dillon		195/4	Henry Nicholas Nicholls
1898	Wesley William Wilson		1955	Stephen William Farrington
1900	Edward Glover		1956	Cornelius John Buckley
1902	John Henry Ryan		1957	Edward Joseph Francis Bourke
1904	Robert Cochrane		1958	Jeremiah Augustine O'Riordan
1906	William Ross		1959	Vernon Dunbavin Harty
1907	Joseph Henry Moore		1960	Jeremiah Gerard Coffey
1909	George Murray Ross		1961	Thomas Aloysius Simington
1911	Peter Chalmers Cowan		1962	Thomas Joseph O'Connor
1913	William Garibaldi Collen		1963	John Lane
1915	Mark Ruddle		1964	Thomas Kelly
1917	Walter Elsworthy Lilly		1965	Richard Ernest Cross
1918	John Ousley Bonsall Moynan		1966	Daniel Herlihy
1919	Patrick Hartnett McCarthy		1967	Patrick Raftery
1920	Francis Bergin		1968	James Clement Ignatius Dooge

Chapter Two
The Presidential Addresses

The long tradition of the presidential address began on the evening of 9 December 1856 at 41 Upper Sackville (O'Connell) Street in Dublin. It had not previously been the custom to deliver an address from the chair, but there were precedents at the time in other professional societies. The addresses afforded a convenient opportunity to review the work accomplished during the preceding period and to add some thoughts for the future. The speaker on this occasion was **George Willoughby Hemans**, an experienced railway engineer who, as chief engineer to the Midland Great Western Railway, had a few years earlier been responsible for completing the main line from Dublin to Galway, and who was to serve two years in office as president. Earlier presidents, such as Burgoyne, Radcliff and Jones, who had headed up the Board of Public Works (Ireland), later to become known as the Office of Public Works (OPW), did not make a formal address to the relatively small number of members, and Richard Griffith reserved his address for the second occasion on which he served as president. These early presidents, together with the support of the county surveyors, had been responsible for the foundation of the Institution.

In his address, Willoughby stressed the need to organise the civil engineering profession in Ireland and to test candidates for admission to the Institution (there was at the time no law preventing any person practising as a civil engineer and placing the letters C.E. after their name. He alluded to the deficiency in papers being presented, particularly as there were so many important engineering works being undertaken in the country, such as the Boyne Viaduct. Willoughby followed the example of Robert Stephenson, the then president of the Institution of Civil Engineers (ICE), in making railways the theme of his address.

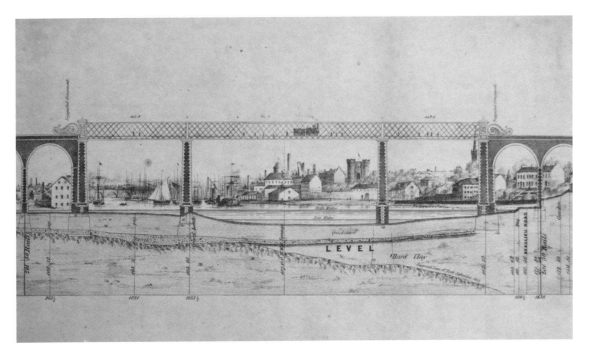

Bridge section of the Boyne Viaduct at Drogheda 1855

The cost of constructing lines of railway in Ireland at about £15,000 per mile compared with £38,000 in England, but over half the Irish railways were single line, and there were fewer major engineering challenges. Willoughby was a firm supporter of proposals to extend Ireland's railway network to less populated agricultural districts by the use of Baronial guarantees of the capital advanced. These consisted of rates-in-aid levied on the counties or baronies traversed in order to secure a minimum dividend to the shareholders in the undertaking. He concluded with a brief review of many of the major engineering projects being undertaken in the 1850s, including the trans-atlantic telegraph cable, involving surveys of the seabed between St John's, Newfoundland and Valentia Island, the building of the *Great Eastern* by Brunel, and the introduction of the Bessemer process for the treatment of iron. Other projects at the time included Brunel's Saltash bridge, the Crumlin viaduct in Wales, and the Victoria tubular bridge in Canada.

Willoughby was succeeded as president in 1859 by **Michael Bernard Mullins**. Possibly as a result of Mullins having a fine private library and archive on which to draw inspiration, his presidential address turned out to be so extensive as to require three evenings (8 November 1859, 13 March and 22 May 1860) to be read to a very patient and attentive audience. It consisted of a history of civil engineering in Ireland and filled no less than 186 pages of the published Transactions. Mullin's successor as president, Richard Griffith, hoped 'that it would provide a sound foundation for future historical accounts of works in Ireland'.

Acknowledging the importance of the port of Dublin as the major entry point for corn, flour and coal, and the establishment in 1707 of the Ballast Board, Mullins began by detailing the early attempts to make the river-based port more accessible. He then went on to discuss inland navigations, both summit-level canals, such as the Grand and Royal canals, and improved river navigations, such as the Shannon. Much has been written subsequently about the history of the inland navigations of Ireland, but Mullin's paper has always provided a useful starting point for engineering historians.

The principal harbours of Ireland were similarly treated, and his address contained much valuable information on the methods used in their construction and adaptation to meet the needs of larger vessels.

Accompanying a brief historical account of road building up to the establishment of the posts of county surveyor, Mullins gives some details of bridges in the Dublin area, but had not had an opportunity to collect information on the many other bridges throughout the country, a deficiency that has only in recent years been significantly reduced by O'Keeffe, Gould, Cox, Donald, and others.

Mullins then turned his attention to chronicling the major works carried out by the Shannon Commissioners to improve the navigation of the River Shannon and the associated arterial drainage schemes. He suggested that the late seventeenth century interest in land drainage may have stemmed from observing the work of the Dutch engineer, Vermyden, in the draining of the Fens in eastern England.

Having made some remarks about water and steam power, Mullins continued his extensive discourse by discussing the progress made in Ireland with railways, commencing by noting that the Dublin-Kingstown line was promoted not long after the opening of the Liverpool-Manchester railway. He felt that the atlas that accompanied the second report of the Railway Commissioners was 'an admirable example of the manner in which the preliminaries of a general railway system, adapted to the wants of a whole country, should be laid down'. He records that William Dargan, the contractor for the greater portion of the Irish railways, had personally undertaken the risk and cost of the Great Industrial Exhibition, held in Dublin in 1853.

Construction of portion of the Rathdrum Rail Viaduct, county Wicklow c1860

Finally, Mullins made some remarks concerning the status of engineers and increasing specialisation in the profession. He said that 'the construction of railways, although of a routine character generally, has, undoubtedly, by their magnitude and importance, raised the profession to a position it had not previously occupied in public estimation' and observed that
> '...our profession, properly embracing all that concerns mechanical construction, having left architecture to its natural alliance with the fine arts, has been divided into three distinct branches, the first comprising masonry, timber, iron and earth work; the second, mechanical engineering; and the third, mining engineering'. 'I am now obliged to bring to a premature close this imperfect and, somewhat disjointed narrative,...'!

So ended Mullin's marathon presidential address.

When **Richard Griffith** delivered his presidential address on the evening of the 19 February 1861, it was the first occasion on which the Institution members had met in the lecture hall of the newly opened Museum Building in Trinity College Dublin. At the invitation of the then Professor of Civil Engineering, Samuel Downing, the Board of the college had graciously consented to providing the Institution with a temporary home for its activities, including its library and engineering models (which supplemented those of the museum of engineering models provided by the college).

Griffith took the opportunity to put forward his views on the education of civil engineers and in particular the importance of a sound knowledge of geology in support of earth works associated with railway construction, such as cuttings, embankments and tunnels, and the identification and procurement of building materials. He had great praise for Bartholemew and Humphrey Lloyd for their promotion of the mechanical sciences and the timely foundation of the Trinity School of Engineering in 1841. Griffith noted that the school supported his contention that 'civil engineering (should not be) limited to the bare exercise of formal, mathematical, and mechanical appliances...', but (should) combine theoretical and practical instruction requisite for the profession of civil engineering', and that this should include geology, mineralogy and chemistry applied to the arts of construction. He continued by tracing the successive introduction of the Licence, Diploma, and Masters Degree in Civil Engineering and observed that by then over one hundred graduates from the Trinity course had been launched on engineering careers, both in Ireland and abroad, representing many future members of the Institution. It is surprising that Griffith did not allude to the fact that civil engineering courses had also been offered by the Queen's colleges in Cork, Galway and Belfast, from around 1845.

Turning to the activities of the Institution, Griffith referred to 'the late apathetic silence of our members' when it came to contributing papers to the Transactions, and was disappointed that the challenge set by Michael Mullins had not been met. Griffith stressed that even short papers were valuable if they led to discussion amongst the members as 'the benefits derived from our Institution are, in a great measure, produced by the different views on each subject which are elicited by freedom of discussion…' He looked to the 'skill and intelligence of the county surveyors', to whom 'we are indebted for the practical solution of many perplexing problems' and 'who have so many opportunities of adding to our knowledge', to 'arise from their lethargy' and contribute papers to the Transactions.

A large part of Griffith's address was devoted to a review of the development of the Irish railway system since the report and recommendations of the Irish Railway Commission some twenty-five years previously. His remarks were influenced by George Willoughby's paper to the ICE in 1858 'On the Railway System in Ireland, the Government Aid afforded, and the Nature and Results of County Guarantees'. Although the main lines to the south and south-west of the country had been built along the route corridors suggested by the geology represented in Griffith's geological map of Ireland, and subsequently recommended by the commissioners, those to the north had followed a coastal route, rather than an inland route. Griffith repeated the need for government aid in the form of baronial guarantees for lines to the west (aid which did come later in the century for some of the less profitable lines).

In 1845, the Institution had been fortunate to have the services of **Robert Mallet** as Secretary at a time when the Institution was placed on a firm footing with the introduction of By-Laws (largely drafted by Mallet) and the publication of Transactions. By the time of his presidency in 1866 at the age of 56, Mallet, together with his father in the iron-founding business of J & R Mallet in Dublin, had contributed much to the development of Ireland's infrastructure, including the design and erection of iron bridges, station roofs, signaling and permanent-way equipment, and much else associated with the railways from their early beginnings in the 1830s. Apart from his engineering and scientific research and experimentation,

Mallet was a noted linguist and writer on technical matters; and this is evident in his presidential address which, whilst not equal to Mullin's epic address, did nevertheless occupy some 54 pages of the Transactions and covered a wide range of topics deemed of importance and interest to the members.

Shannon Road Bridge at Athlone with Mallet opening span

Due to the economic downturn in the country occasioned, not only by the fact that the main lines of railway had largely been completed, but that iron foundry work could not compete with English products, due to the fact that iron ore and coal had to be imported, Mallet moved to London. Here he continued his scientific research and writing, and he was conscious of the effect this would have on his ability to fulfill his role as president. Mallet was the first of number of presidents who chose to trace the early history of the Institution. In 1844, Burgoyne, the first president, sought Mallet's advice as to what to do about the Society (as it then was) and Mallet's replied that 'in no event should the Institution be permitted to die'. This early history has been traced in Chapter One.

Commenting on the introduction in the late 1840s of the government policy of '*laissez faire*', whereby 'those who had been reduced to helplessness were to help themselves', Mallet opined that it had been 'a disastrous policy, which (had) reduced the functions of the Board of Public Works of Ireland almost to a cipher (one that has no weight, worth, or influence: a nonentity), and opposed to which the sonorous nonsense of Anglo-Saxon enterprise and the wisdom of self-reliance are no refutation'. The results of the policy were to diminish in Ireland the avenues of employment of engineers and others associated with the profession. Mallet mentioned the isolation of the profession in Ireland and felt that the Institution members should become more informed of the experiences and best practices abroad.

Mallet supported the awarding, by the schools of engineering, of diplomas, in order to keep at bay 'the swarm of uneducated or incompetent pretenders to the title Civil Engineer...from injuring the reputation of the profession'. He, nevertheless, recognized that a diploma on its own simply testified that the holder had diligently pursued a course of special study, and did not make him an engineer, but fitted him to become one, and that there was a radical difference between a diploma and a 'licence to practise' engineering.

Having indicated under a number of headings what he felt engineers should be taught, Mallet proceeded to review some of the great engineering undertakings recently completed or in progress in the world. He noted that, since the commencement of railways as a mode of travelling, some one billion pounds had been expended on some 70,000 miles of lines throughout the world. Naturally enough, Mallet also used the occasion to discuss advances in the design of iron ships and ordnance, which had incorporated a number of his ideas, and the properties of steel, the use of which was increasing following the introduction of the Bessemer conversion process. He noted that oil springs had been discovered in America and that it was envisaged that oil might one day replace coal in marine boilers.

Mallet felt that 'the social circumstances of Ireland (were) exceptional, and particularly unfavorable in some respects for the development of engineering or industrial works', and cited as an example that 'from one end of Ireland to the other scarcely a good brick can be purchased'. The cessation of the arterial drainage works had been, in his opinion, suicidal, and urged that they be resumed immediately at government expense.

The next president, **William Anderson**, was the first purely mechanical engineer to serve as president. By this time (1867) he had moved back to England to join the firm of Easton & Amos in London. He confined his observations to manufacturing and mechanical sciences rather than the broad and comprehensive surveys of civil engineering that had been favoured by his predecessors. Quoting Bacon, who said that *'reading makes a full man, writing an exact man, and speaking a ready man'*, Anderson felt that an engineer should possess all these qualities, and he appealed to the younger members of the Institution not to neglect the opportunities afforded by general meetings of acquiring the art of public speaking and writing.

He provided descriptions of a selection of industrial processes that he considered of importance, in particular the manufacture of paper and the artificial drying and consolidation of peat as a fuel. This latter process had been successfully developed in Ireland by Charles Hodgson, but had met with financial difficulties and had ceased operation. Anderson, together with Charles Philip Cotton, had spent a month in 1865 investigating the production facility and had reported that the cost of producing the compressed peat had shown a profit of 3s per ton and would have continued to be a viable operation had the company started with sufficient capital. He noted that iron shipbuilding had been established in Belfast, Dublin, Drogheda, Cork and Waterford, and that Dublin was fast becoming a deepwater port.

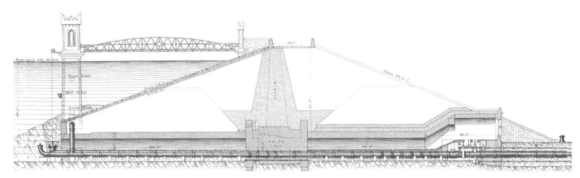

Section through the earth dam near Roundwood, part of the Vartry Scheme

Anderson took the opportunity to discuss the contentious issue of the claim of Richard Hassard to have been the designer of the Dublin Corporation Water Supply scheme based on the Vartry River in County Wicklow, rather than the City Engineer, Parke Neville. It was generally considered that Hassard, the originator and projector of the scheme, as set down in his paper to the Institution in 1861, had been most unfairly deprived of the honour and emoluments due to him. Parke Neville did later acknowledge that Hassard's scheme had been adopted with some trifling modifications, but that the detailed design and supervision of the scheme had been carried out by him as the Corporation's Engineer-in Chief. Anderson provides an interesting summary of the events leading to the near collapse of the earth dam near Roundwood and the remedial works necessary to make the structure safe. He concluded that 'the citizens of Dublin will have the satisfaction of knowing that they have paid a good many thousand pounds for the professional education, so far as hydraulics are concerned, of the Water Works Committee and its officers'.

Following on was the Commissioner of Valuation, **John Ball Greene**. He apologized for not being able to attend many of the papers during the period of his presidency, but, nevertheless, decided to devote a major part of his address to a review of the papers, beginning with that given by a Dr Kidd, a Fellow of the Royal College of Surgeons, on a system for overcoming the problem of frozen pipes in domestic dwellings that he had devised. Essentially, his idea was to introduce a number of small cisterns and stop-cocks at various locations in a system such that the most vulnerable sections of pipe were empty of water when not in use, thus preventing water freezing in the pipes. Samuel Geoghegan in a following paper presented much the same idea, but suggested that such a system could be made 'self-acting'. A past-president, William Anderson, wrote on 'domestic water works', i.e. individual water schemes for large houses, employing the Poncelet water wheel for low heads or turbines to raise the water to a sufficient head to ensure an adequate supply.

A paper describing the Bann reservoir at Lough Island Reavy, and the repairs carried out to counteract a leak in the main reservoir, prompted Ball Greene to refer to the earlier similar problems at the dam of the Dublin water works at Roundwood. He also used the opportunity to refer to the part played in the scheme by his brother Henry, who had been one of the principal contractors on the two-mile long aqueduct tunnel from below the dam towards Newtownmountkennedy. As a result of the difficult tunneling conditions occasioned by the hardness of the Pre-Cambrian rock, Henry Greene lost everything (about £31,000) on the contract. Added to this, on 14 October 1868 he tragically fell from a wall in Glendalough and five days later died from a brain hemorrhage, leaving a wife and fourteen children. A trust fund was set up by generous public subscription that saved his family from certain penury.

The usefulness of the Transactions as a record of engineering achievements in Ireland is exemplified by the detailed paper by William Strype, the manager of the Grendon ironworks in Drogheda, on the construction of the wrought-iron latticed girder road bridge over the River Boyne near the site of the Battle of the Boyne to replace an earlier timber structure.

Before ending with a review of some engineering achievements abroad, including the Suez Canal and the Mont-Cenis tunnel, Ball Greene referred to the then recession (1870) in Ireland and continued 'Should the legislative expedients now in contemplation hasten the advent of a more prosperous period, the unhappy circumstances hitherto surrounding our unfortunate country may be dispelled, and replaced by a corresponding increase of the improvements and projects in which we have, in comparison to other countries, been so long deficient; and should Government be successful in effecting a satisfactory arrangement in connexion with the occupation of land in Ireland, we may reasonably hope to participate in the benefits arising from the restoration of public confidence and enterprise likely to ensure'. Some things never change! As things turned out, the Landlord and Tenant (Ireland) Act of 1870 signaled the beginning of the 'Land War' and the activities of the Land League led by Michael Davitt.

Alexandra Basin, Port of Dublin (Dublin Port Archives)

By the time that **Bindon Blood Stoney** rose to give his presidential address on 12 January 1872, the Church in Ireland, of which Stoney was an active member of Donnybrook parish, had been disestablished (the Irish Church Act of 1869 separated the church from the state, abolished the payment of tithes, the church from then on being known as the Church of Ireland). Regarding the status of the profession, Stoney regretted that 'anyone who wishes to claim the title civil engineer may do so without check or remonstrance'. He pointed out that there were now a number of routes to qualification: a) a diploma or degree of a recognized school of engineering; b) a pupilage for a minimum of three years; and c) membership of the Institution of Civil Engineers.

Stoney stressed the need for a suitable combination of scientific training and practical knowledge and stated that:
'hitherto, it had been too much the practice to consider that ordinary mechanical tastes are enough to stamp a boy as being suited for the profession, and fond parents are sometimes apt to think that, because their son prefers making rabbit hutches, boats, or such-like toys to learning his lessons diligently, that therefore his tastes mark him out as a future Rennie or a Stephenson'. 'There can be no greater mistake than this, and in no profession perhaps, now that competition has become excessive, is there so little prospect that men who are mentally deficient, or who are averse to intellectual labour, will succeed in obtaining a high position'. As to what methods of student assessment were most appropriate in the colleges, he cited the example of the École Centrale in Paris that employed a system of continuous assessment and used third-year marks in the final assessment in the fourth year, a system that became widely used.

Regarding the future prospects for Ireland, Stoney favoured a three-foot gauge for an expanded railway network, due to the country not being sufficiently industrialised. He felt that a three-foot gauge should have been adopted originally and a frequent service of light trains introduced to encourage rail travel.

Being chief engineer at Dublin port, Stoney noted that 'few circumstances afford stronger evidence of the increasing prosperity of this country than the development of traffic; and this is shown to a marked degree by the increased tonnage and demand for improvement in some of our principal ports'; the tonnage at Belfast and Dublin had doubled in the space of two decades. He noted that, with the ending of the Franco-Prussian war, trade had picked up, especially so since the expiration of the Bessemer patents provided a stimulus to iron manufacture. On the subject of sanitary engineering, he advocated the establishment of drainage boards for each of the main river systems.

Charles Philip Cotton studied previous presidential addresses in order 'avoid repetition', presumably an approach welcomed by the members present in 1873! It was now nearly fifteen years since Mullins had traced the history of civil engineering in Ireland in his monumental address, and Cotton was keen to bring it up to date. Cotton's career had taken him from railways to public health and he naturally began by describing the lines of railways that had been completed since 1860. These included lines to Westport and Sligo opening up the west and the completion of the east coast route to Wexford. His experience led him to publish in 1861 *A Manual of Railway Engineering in Ireland*. He noted that there had been no less than twenty-six different plans for uniting the Dublin railway termini (up to 1874). These included the earliest proposal put forward by Vignoles in 1837 to connect the terminus of the Dublin-Kingstown railway at Westland Row with the planned terminus of the main line to Cork. This essentially contemplated a railway supported on girders atop a double row of masonry columns along the south quays of the River Liffey, one row being at the edge of the footpath, the other in the river. His plan included provision for a catch-sewer along the length of the colonnade. Over twenty years were to pass before LeFanu proposed a scheme to connect all the railways, commencing near the mouth of the River Dodder and running through Rathgar to the GSWR at Kingsbridge and thence across the Liffey and under the Phoenix Park to what became known as Liffey Junction and thence to the quayside at North Wall. The section from Kingsbridge (now Heuston) station to the North Wall was later built as LeFanu had envisaged. All the other schemes were based on one or other of the two original proposals. Some fifteen years later, Westland Row (Pearse) and Amiens Street (Connolly) were to be connected directly via the Liffey Viaduct.

Cotton then turned his attention to dock and harbour projects and cited Barton's work at Greenore, Stoney's at Dublin Port, and that of Price on movable bridges, in particular at Spencer Dock in Dublin. All chose to describe their works in papers to the Institution of Civil Engineers (ICE) rather than the Dublin institution, a fact that caused Cotton to express his regret 'that our archives are not the repository of the history and details of such Irish engineering work'. However, he added that 'the author of a unique work is right to go to the very highest authorities for criticism, and, beyond doubt, praise.' He mentioned the new harbour then being created at Rosslare, a work in which Portland cement concrete was almost exclusively used. To avoid interfering with the shore currents, the design of the harbour involved the construction of a curving pier in deep water connected to the shore by an open viaduct consisting of iron girders supported on concrete columns.

Cotton concluded his address by referring to the move of the Institution from Trinity College Dublin to rooms in the upper part of 136 St Stephens Green. As was mentioned in Chapter One, this had been made possible as a result of the Mullins bequest of £1,200 and by an increase in membership subscriptions. In his will, Mullins also bequeathed his considerable library of engineering and architectural books to form the library of the Institution. Cotton was keen to increase the membership, particular the number of Associates.

In a rather lengthy address, delivered on 16 February 1876, **Alexander McDonnell** chose to review engineering progress and some of the major projects recently completed or in progress. These were mostly outside Ireland, although he was quick to remark that the prosperity of the country had been increased by the work of engineers. Prior to the generation and transmission of electrical power, hydraulic power, the transmission of power by wire ropes, or by compressed air, represented the latest technology.

On the cleaning of streets in cities, he was moved to cite the case of Dublin, where he remarked that 'it would be difficult to find a sight more annoying to an engineer, whose chief business is to economise labour, than that of the men who are scattered in twos and threes through Dublin, sometimes working, sometimes wandering with their brushes on their shoulders, and sometimes, when they have nothing else to do, sweeping the mud into heaps'. He felt that

> 'it is ridiculous to expect that men will work hard to sweep mud off the streets when they know perfectly well that it will be left there, and that in a week or ten days, they will have the chance of sweeping it up again.'

Describing the situation in the Liberties of Dublin of a few years back, when around 2,000 houses were occupied by a population of some 40,000, McDonnell wrote that the houses were built in narrow lanes and alleys, often in blocks without any yards, and continued 'The cesspool, where there is one, is filthy, and not only fills the air about with stench, but saturates the ground with sewage'. McDonnell found that 'it was impossible to conceive a district which could be more benefited by an improved system of drainage, and by new houses built under the Artizans Dwelling Act'. He added 'It must be admitted that the state of Dublin, both as to drainage and the state of the streets, is not a credit to the country'.

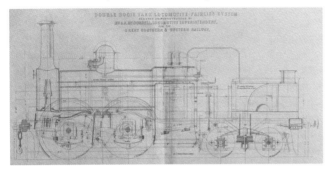

As was to be expected, a large part of McDonnell's address was devoted to an analysis of the operation of railways, both in Ireland and, by comparison, with those in Britain and further afield, but he finished with an exhortation to the young engineer preparing to commence their duties as follows:

'It is, of course, necessary to become well grounded in mathematics, mechanics, and chemistry, which are now so necessary for all well-educated engineers. The better your knowledge of these, the more you will use your knowledge - I might almost say without your knowing it. Pay particular attention to geometry of all kinds, to the geometry of three dimensions and geometrical mechanics. It will assist you in designing, and in mental calculations, to which I would advise you to become habituated. Learn French, and, if possible, German. Learn a little book-keeping, and become well acquainted with accounts, and the use they can be put to when properly kept. Without a knowledge of accounts, you will never be able to carry out large works with economy, or ever become thoroughly acquainted with the cost of work, which is of the first importance to an engineer. As soon as you have an opportunity, learn the cost of materials of all kinds, and the uses, different qualities of materials can be put to with advantage. If you undertake the management of workmen, study their peculiarities. The workmen of no nation work hard if left to themselves. You will get better results by organising their work well than by trying to force them to work hard by unnecessary severity. Enforce always strict discipline with strict justice. Remember that men must, of necessity, be led by some one, and let them feel that you are naturally the person most interested in their welfare, and that if they have anything to complain of, you are the first person they should come to, to have their grievance remedied. Learn to know good from bad work, and a good from a bad workman. Become, as soon as possible, good men of business, and depend on your own talents and exertions for success. Above all, do not attempt too much, but whatever you undertake to do, do well'.

Having been elected to serve a second year as President, **Robert Manning** found himself in a pivotal role as the principal person named in a Royal Charter granted in 1878 to the Institution of Civil Engineers of Ireland by Queen Victoria, thus creating a body corporate of engineers [see Chapter One]. As the Institution had entered a new phase of its existence, Manning felt it was opportune to take a retrospective look at the rise and progress of civil engineering in Ireland, the relationship of the Institution to such progress, and the future prospects for both. In his review, he noted that between 1835 and 1878, 2,157 miles of railway had been constructed in Ireland at a cost of £30 million, and that most rivers had been improved and over 300,000 acres of land freed from flooding. Apart from the 68 fishery piers mentioned in the First report of the Fishery Commissioners, a further 100 had been built or improved (up to 1878).

Manning felt that there should be less dependence on British engineers and voiced his opinion that 'We have Irishmen – educated in Ireland – who are capable of designing and executing whatever works the progress of the country may demand'. He would have been saddened to acknowledge the dependence on foreign expertise for the success of such projects as the Shannon Scheme in the 1920s. Manning was one of the first engineers to use Portland cement in Ireland, in 1850 for the underpinning of a bridge in the tideway of the River Glyde. Robert Manning is, of course, best remembered for his research on the flow of water in pipes and open channels, his empirical formula for open channel flow (shown here in its original form as communicated to the Institution in 1889) still being in use today.

$$V = C \sqrt{S\,g.} \left[R^{\frac{1}{2}} + \frac{0 \cdot 22}{m^{\frac{1}{4}}} (R - 0 \cdot 15\,m) \right]$$

John Bailey was a mechanical engineer who came to Dublin in 1865. He prefaced his presidential address in 1880 with short biographies of four eminent personages, then recently deceased: Sir Richard Griffith, Sir Thomas Larcon, Marcus Harty, and Sir John Macneill. He mentioned that Harty had contributed a paper to the Transactions on the reconstruction of the Malahide Viaduct, a paper that was to contribute to the historical background information found useful following the collapse of two of the spans of the viaduct in 2009. It was noted that John Macneill had lodged his reports and correspondence (1827-1844) in the Institution's library. These now form part of the Institution's archives that are maintained by the Irish Architectural Archive. He was noted as not having been a member of the Institution, but recent research in the membership application records has revealed that Sir John was indeed a member.

During Bailey's term of office, interest in the lease of 35 Dawson Street in the city was purchased and it was planned to erect a spacious theatre or lecture room at the rear of the main building. On completion of the building work, the President said, we shall 'bid farewell to the noble college (TCD) within whose walls we are now assembled, and which has so generously sheltered us for so many years, standing our friend when most needed, and thus exemplifying the old adage, *"a friend in need is a friend in deed"*. As a result of the Mullin's bequest, the first two Mullins Gold Medals were awarded for best papers, and it was hoped that the medals would have the effect of stimulating the members of the Institution to come forward with papers in greater abundance, and that they would not forget that the Institution 'claims a paper from them before that of any other similar institution or constituted body'.

In his address, Bailey dwelt mainly on aspects of mechanical science, but observed that civil and mechanical engineering were bound together and necessary each to each. He continued
> 'Vast as has been the work of the purely civil engineer, I think you will agree with me that the achievements of his mechanical brother have fully equalled them, and that the latter by his countless inventions and wonderful appliances has enabled the former to bring those very works to a successful and happy issue'.

He recommended young civil engineers spending time in mechanical engineering workshops. Bailey advocated a system of steam tramways for Dublin, but the advent of electricity within a decade was to see this new form of motive power utilised to the full on the streets of many cities, including Dublin, where a merging of a number of companies to form the Dublin United Tramways Co. ensured that an efficient integrated public transport system was for the first time offered to its citizens.

Bailey also commented on the 'Battle of the Brakes'. As soon as it was recognised in England that a continuous and automatic braking system on trains must sooner or later become a requirement, numerous competing patents were taken out purporting to accomplish the desired object, amongst the most popular being that developed by Westinghouse. The end of 1880 saw the Tay Bridge disaster and the ensuing Commission of Enquiry, which, in like manner to the 1847 enquiry into the Dee Bridge collapse, led to fundamental changes in the design of railway bridges, in particular the cessation of the use of cast iron and the promotion of studies of the effect of wind loading on structures.

The Dublin City Engineer, **Parke Neville**, was the next to take the chair, which he did following a five-month illness that had prevented him from attending meetings. In his presidential address, delivered on 1 March 1882, he began by alluding to the great benefits that had accrued from the granting of the Royal Charter to the Institution. He felt that membership of the Institution 'should give some safety to the

Intake tower and access bridge at Roundwood Reservoir, county Wicklow

public against the employment of the quacks of all sorts, who now call themselves engineers, without any qualification whatsoever'.

Not surprisingly, Neville chose to talk about the water supply to Dublin, eighteen years having elapsed since the scheme was completed. He gave a detailed description of the Vartry scheme, from its inception to completion. The original scheme was designed to supply the city and the townships of Pembroke and Blackrock (Rathmines and Rathgar township having its own supply). During the decade following completion, supplies were also provided for the townships of Clontarf, Kilmainham, Kingstown, Dalkey, Killiney and Ballybrack, and Bray. Speaking of the future need to expand the scheme to develop the full potential of the Vartry River catchment, Neville noted that this could be done by the construction of a second reservoir to the north of Lough Vartry and that a second parallel main would then need to be laid from the end of the aqueduct tunnel to the distribution reservoirs at Stillorgan (a second service reservoir was then under construction). The second reservoir at Roundwood was not completed until 1925, some forty years later. In his opinion, water should not be made available by the city authorities for motive power and should be limited to domestic use and ordinary manufacturing purposes, such as brewing.

It was then the turn of a railway engineer to be elected president. **William Hemingway Mills**, the Chief Engineer of the Great Northern Railways (Ireland) delivered his inaugural address in November 1883 when he prefaced his remarks by reminding his audience that 'As engineers we have every reason to be proud of our profession which takes such a prominent part in contributing to the comfort, convenience, and civilisation of the world'.

Rather than talk solely about railways, Mills chose to make his address much more general and to describe some of the major engineering projects being undertaken or recently completed. Mentioning the Forth Bridge (then building), Mills rather presciently noted that such a large steel structure in a marine environment would result in heavy annual expenditure on inspection, painting and general maintenance. How right he was! He spent some time speaking about the introduction of steel rails and the need to think

21

ahead to what would replace timber sleepers when we had cut down all the available forests. He concluded that

> 'unless some great change take place, and planting be carried out on a grand scale, we must look to some other material for this important item of our permanent way'. Mills ended by stressing the importance of engineering to the human condition. 'The records of past experience, and the knowledge gained from scientific research, give us (in 1882) immense advantage over our predecessors, who had few practical examples to refer to…' 'Still, with all this information before him, the engineer cannot relax the thorough consideration that is incumbent upon him, for, in assuming the duties of his profession, the engineer must fully understand that it is second to none in importance'. 'The most important duty is where there is the greatest responsibility, and there is no greater responsibility than the care of human life'.

John Aspinall had already served for over a year as president before giving his address in November, 1885. A mechanical engineer, Aspinall drew the attention of members to the development over the previous five years of steam power for working tramways, due he said, in great measure to the greater perfection in the construction of the engines,

> 'which have been made not only less noisy and so less likely to frighten horses, but also to the fact of their being made with working parts having larger bearing surfaces, which tend to keep the engine out of the repair shop for a longer period, thus reducing the cost of maintenance'.

Other forms of tramway at the time used engines running on compressed air, and, in a few cases, operated by cable systems. It was to be a few more years before electric motive power was developed to the extent of replacing all other systems of powering tramways.

Regarding the question of the fueling of steam locomotives, and the dearth of native supplies of coal, Aspinall was often asked why turf was not used on the railways in Ireland. He pointed out that it had been tried repeatedly, but without any real success. He went on to summarise the situation thuswise:

> 'The supply of cut turf is too dependent upon good seasons to be relied upon, and even if it could be supplied with regularity, it will not stand the punishment in the fire-box to which ordinary coal is subjected when burning in a locomotive drawing a heavy train'. 'If however the turf is compressed and well dried it makes a fairly good fuel, but no efforts to produce it in this form have been commercially successful, and the history of the many peat works which have been started would fill many pages'.

He alluded to a paper by Alexander McDonnell read before the Institution in 1874, which went fully into the subject.

On the subject of electric light taking over from gas lighting, Aspinall had this to say:

> 'We have not as yet arrived at the time when anyone can, with reasonable economy, light his house with incandescent lamps at a price which will compare favourably with ordinary illuminants; and unless the ordinary householder is capable of understanding and attending to an engine and dynamo, he must yet wait before he can gain the undoubted advantage of the electric light at a reasonable cost'.

Aspinall noted that

> 'Those who have had anything to do with the creation of electric lighting plant cannot help being struck with the number of German workmen who are employed in connection with this work, and it is no doubt due to their possessing a greater knowledge of electricity and chemistry which our own workmen have not the same facilities for obtaining; and this fact is but one of many which may be brought forward to illustrate the necessity of the establishment of technical and scientific schools in our large towns'. (Two years later, in 1887, Kevin Street Technical School was founded in Dublin).

In conclusion, Aspinall made reference, not only to the novelty in the design of the Forth Rail Bridge, but the novelty in almost every machine used in its construction, such as special tools for bending, drilling, planning and riveting the plates, and that 'No greater or better illustration can be found to show that in these days mechanical genius must be found hand in hand with scientific research if great results are to follow'.

John (later Sir John) Purser Griffith assumed the presidency in 1887. In his address, he began by tracing the early history of the Institution, recalling its objectives, as put forward by Burgoyne in 1835. Since that time, architects had gone their own separate ways and were now no longer eligible for membership. Griffith reminded his audience that,

> 'although the work constructed by the civil engineer may be destitute of architectural ornament, yet no architectural work can be considered worthy of the name unless designed on sound structural principles'. 'Architecture, therefore, seems to me a department of civil engineering, and I cannot but think it a matter of regret that it should have become a separate profession'.

Efforts were being made at the time to obtain an Act of Parliament for the registration of architects, civil engineers and surveyors, the object being to provide the public with a means of distinguishing qualified from unqualified practitioners. The Council of the ICE had recently warned that the letters CE were not based on any qualification and were calculated to mislead. Griffith urged the abandonment of CE and the use instead of MICEI or AICEI.

He felt that the profession had an important role to play in the testing and accreditation of engineering courses offered by the schools of engineering in the universities. He believed that the ICEI was one of the first professional bodies to acknowledge the curriculum of certain engineering schools as part training of an engineer.

Griffith saw the Institution as assisting with the introduction of standard tests for materials used in construction, and proof tests for completed structures, rather than the plethora of different tests and specifications then current. He viewed with satisfaction the formation of engineering laboratories where such standard tests could be undertaken.

His address then took the form of a retrospective look at the progress of engineering during the previous fifty years. Railway safety had improved twelve-fold, thanks to the maintenance of rolling stock, signalling, the block system, and continuous brakes, but Griffith felt that there was still a pressing need for better communication between the passengers and the guard.

He believed it to be an important duty of engineers to unite efficiency and economy in all their designs and that 'failure of works on economic grounds brings discredit on our profession, second only to failures from structural defects'. Of the telephone, he remarked that 'but for the convenience of rapid communication, it would be expelled from many an office as an unqualified nuisance and rude companion'.

In view of his career as a dock and harbours engineer, Griffith naturally provided many details of work at London, Liverpool and other ports, in particular Dublin, where he referred to the introduction of Portland cement as having revolutionised the construction of marine works. Having dealt with progress in water supply and sewage disposal, he described the major works required to complete the Panama, Corinth, and Manchester ship canals. The other major projects in train at the time were the Tay and Forth rail bridges.

Firth of Forth Rail Bridge

Griffith concluded by explaining that he had dwelt at some length with the history of the profession in order to stimulate his listeners to assist in making the future (of the profession) worthy of its past. He went on:

> 'Do not be discouraged by anything you have heard this evening, but if your heart is in your work go on fearlessly. To ensure success you must, however, be enthusiasts. It will not do to enter the profession merely because it appears a gentlemanly occupation. If you join it from such motives you may expect failure. Remember you are taking your commission in the constructive army of the world, and, as in the military service, much of your success will turn on your becoming distinguished as leaders of men. Study engineering biography, and you will find that the most eminent engineers have been those who possessed the power of gathering around them men thoroughly devoted and loyal to their interests. Let it be your endeavour, in whatever position you are placed, to surround yourselves with such a body-guard. Whether as assistant or executive engineers, strive to make yourselves respected by those under you for your personal as well as your professional worth. Let them learn to look up to you as their counsellor and leader in all cases of difficulty. The progress of engineering science is now so rapid that it is no easy matter for men whose days are fully occupied with professional engagements, to keep themselves informed of the advances which are being made. Yet this is necessary, and you must not shrink from the labour it entails. To ensure a successful career you must keep in touch with the most recent discoveries and improvements in the art of construction, so as to be able to utilise them as occasion may require. If possible, travel, and visit engineering works of importance; by doing so your views will be enlarged and local prejudices corrected, while a wider personal acquaintance with engineers will increase your interest in, and attachment to, the honourable profession which you have joined'.

The next president, **Spencer Harty**, chose as the subject of his address the recent progress in the municipal engineering of Dublin. His address in 1889 covered the housing of the working classes, sewers, plumbing (wherein he promoted restricting sanitary work to qualified licensed plumbers), new streets, water works and fire service, street making and paving, street cleansing, electric lighting, and cattle market and abattoir.

Throughout his address there is a sense of progress and improvement. However, the limited capital resources at the time restricted progress. Although Harty said that Dublin's streets would bear comparison with the best of English cities, there is a sense that Dublin's facilities did not.

He cited Dublin and Glasgow as among the worst cities in the UK with regard to sewer outfalls. As Borough Surveyor and Waterworks Engineer, Spencer Harty would later oversee the construction of the long-awaited interceptor sewers, which would lead to a marked improvement in Dublin's sewerage system.

In discussing the city's water supply Harty gave unfortunate hostage to fortune. In praising the then new Vartry system he described it as providing an "unlimited supply". This remark was based on the fact that the Vartry reservoir had succeeded in providing a steady supply since its associated scheme had been completed in 1867, including the year 1887 when rainfall had been particularly low. However, in 1893, three years after his address, there was a severe drought that ultimately led to the planning for a second reservoir at Roundwood (completed 1925).

Harty's address traced the history of Dublin's municipal services and painted a vivid picture of the poor housing conditions of Dublin's working class, 'upward of 27 (persons) depending upon one miserable closet fixed up in an open court'. Harty gave much space to The Housing of the Working Classes Act, 1890 and gave an insight into the work of the Dublin Artisans' Dwelling Company and Dublin Workingman's Suburban Dwellings Company. Harty offered to 'insert in the Transactions, if the council so desires, plans, elevations, and sections of some of what are considered in Ireland to be the class of house best suited for the working classes, together with the cost'.

Within the section on water works Harty included an interesting vignette of the early fire service. Telephonic communication had recently come into play and Harty noted that the time for the fire crew to turn out was forty seconds during the day and a maximum of two minutes at night.

Thomas Pigot gave his presidential address in 1891 in the ICEI's newly built 'fine hall'. Pigot made reference to the former kind hospitality of the Provost and Fellows of Trinity College and the space that had previously been provided in the School of Engineering for ICEI meetings. The topic of Pigot's address was 'the extension of railways and harbours, and the present stated and possible improvement of our inland navigation'.

Pigot's opening remarks remain relevant today: He noted that the general prosperity of the country is of paramount importance in the success of public works. In 1891 the financial state of the country had been improving for ten years. The railways were in a healthy state and the amount of Irish railway capital held in England had diminished with by far the greater part being held in Ireland. The Irish mercantile navy comprised 1,205 vessels of a total of about 241,400 tons at the close of 1889, one fifth of the entire fleet having been built in that year.

The total railway track mileage in July 1891 was 2,676. A great impetus had been given to the construction of railways and tramways by the acts of 1883, 1889 and 1890. Pigot mentioned that railways in remote areas gave employment 'in districts threatened with the periodical distress which even the youngest of us can remember from previous years'. The issue of the appropriate gauge for new lines was then a current topic and Pigot stated that he wouldn't get drawn into the relative merits of broad (5ft 3in) and narrow (3ft) gauges. He gave details of some of the recent railway extensions, but did not think that further extension of the network was likely for some years.

The objective of linking up Dublin's railway termini had been under consideration since the report of the Report of the Railway Commissioners' in 1837 and Pigot made particular mention of the Dublin Junction Railway linking Westland Row and Amien Street stations. He gave a very interesting technical insight into how the railway bridge over Westland Row uses 'counterpoises resting on the extremities of the girder' in order to reduce the central depth of the bridge over the street, thereby giving increased headroom. He also complained about the visual intrusion of the loopline bridge vis-á-vis the Custom House, and was destined not to be the last to do so!

At the time of Pigot's address the Irish fishing industry was still underdeveloped. Although there were 182 fishery harbours plus five royal harbours (Howth, Dublin, Dunmore, Donaghadee and Ardglass) the size of the vessels on the west coast was limited by the inaccessibility of the majority of piers to vessels at low water. Pigot reasoned that now that railways had been or were being laid to Baltimore, Schull, Kenmare, Valentia, Clifden, Achill Sound, Killala and Killybegs, the next step should be the construction of harbours that could accommodate vessels large enough to venture into the open sea at every state of tide.

The second half of Pigot's address covered inland navigation in Ireland. He disagreed with the recent 1882 Report of the Commission appointed to inquire into the system of navigation between Coleraine, Belfast and Limerick. This commission had recommended the sale of the Ulster Canal and saw little benefit in restoring the Ballinamore Canal to complete the navigation from the Shannon through Lough Erne and on to Lough Neagh. Pigot however argued for the re-establishment of the navigation. He noted a recent increase in traffic on the Grand Canal, but this was really against a backdrop of a general decline in the canal trade.

Like many presidents before and later, **John Chaloner Smith** began his address in 1893 by encouraging engineers who were not members of the ICEI to join. His address was really a personal soapbox pitch, but with Ireland's economic prosperity at its heart. The entire focus of his address was on the reduction of railway rates. Chaloner Smith had presented a paper in 1884 on "The practicability of reducing railway rates in Ireland". In 1888 a report of the Royal Commission on Irish Public Works proposed a guarantee scheme, which would have resulted in reduced rates, but this report differed in some respects from Chaloner Smith's proposal. Much of his address concerned the relative taxation burden on Ireland as compared with England, Scotland and Wales. Although interesting, this was somewhat peripheral.

Chaloner Smith praised the efforts of Thomas Drummond who, when he was Under-Secretary for Ireland, formulated the policy that should be pursued by Great Britain towards Ireland in the Report on Irish Railways of 1838. One of the resolutions adopted by Parliament arising from the 1838 report was that Commissioners, subject to the approval of the Treasury, should fix the rates of carriage on the railways, however, nothing arose from this decision, and the various railway companies fixed their own competitive rates.

In 1895, **James Price** followed John Chaloner Smith as president, but died before the expiry of his term of office and hence did not deliver an address. His successor, **James Dillon**, commenced his address with a eulogy to Chaloner Smith. The annual Smith Premium had just been established to recognise his service to the Institution, and was to be awarded to 'deserving authors of papers read before this Institution'. The establishment of the prize was reported to have given Chaloner Smith great satisfaction. Dillon encouraged potential authors and reminded them of both the Smith Premium of books and the Mullin's gold and silver medals.

Dillon's presidential address was delivered on 2 December 1896 in the new headquarters of the ICEI at 35 Dawson Street (which comprised Library, Reading Room and Lecture Hall). The subject of Dillon's address was "the public works that Ireland need for the proper development of her resources, including agriculture, deep sea and inland fisheries, and her great live stock export trade and other matters"
Dillon noted that some European countries had transport rates 100% lower than in Ireland. He noted that these countries are: densely populated, better educated, more contented, better fed and clothed, and have larger balances in their banks. In short this made it difficult to ship produce to England competitively. He then listed the Irish fishing harbours and cited the Royal Commission Report on Irish Public Works 1888, stating that no great development of fisheries could be looked for in the absence of proper sea-going vessels, for which it would be necessary to construct deep harbours.

The ICEI Lecture Hall at 35 Dawson Street, Dublin

Dillon cited the miles of railway track per square mile for Ireland, England and Scotland and concluded that Ireland had not yet got an appropriate proportion. He observed that in general, trains from one Irish company were not allowed to run over another's tracks and that there were gaps in the rail network in Irish cities. Dillon reported the total track milage as 3,173 miles of 5ft 3in gauge and 326 miles of narrow gauge and gave the average cost of construction of railway track in Ireland as £5,000/mile. It was his view that government needed to find cheap money for railways and fisheries. In addition, he saw the need for 'good steady sea boats for the safe and quick export' of live meat.

Dillon then discussed Ireland's inland navigation, which he estimated as having comprised 750 miles at a construction cost of £5,000,000. He recognised the conflict between maintaining navigation and arterial drainage in some locations. In addition, the Royal Canal had problems with the depth of water, partly as a result of poor design and also neglect by the railway company. (The Royal Canal Co. was purchased by the Midland and Great Western Railway in 1845).

Royal Canal at Richmond Harbour

Dillon felt that the government had a duty to restore and maintain the canal system.

On the subject of arterial drainage, Dillon noted that little progress had been made prior to the 1842 Arterial Drainage Act. This Act and the later 1863 Arterial Drainage Act had assisted landed proprietors to carry out drainage works with the aid of Government loans. The procedure was for proprietors to hire engineers and other personnel, with the Board of Works advancing the loans as the works proceeded, but the works were never truly arterial in nature. The first report of the Royal Commission on Irish Public Works (1886) gives the past history of arterial drainage in Ireland. Interestingly arterial drainage was linked to 'improving the climate'.

On railways, Dillon advised trying to keep to the 5ft 3in gauge. He lised the proposed extensions and lines, many of which were carried out, but of these, only the link to Ballina has survived. Dillon stated that, as the government couldn't get parliament to sanction the Royal Commission's proposals, it had therefore been decided to at least support certain branch railways. Dillon credited Arthur Balfour, the then Chief Secretary for Ireland, as the driving force behind the Light Railways Act (1889). This act provided a free grant for construction. There was still a surplus of £500,000 available but Dillon did not believe this would meet Ireland's needs.

Dillon next discussed the public works that were then in progress under government departments but noted that no provision had been made at the time for any new engineering works such as harbours, inland navigation, or new arterial drainage work. He complained stridently that the Royal Commission had insufficient representation of Irish civil engineers and he also commented that private bill legislation was defective and too costly.

On the topic of engineering education, Dillon believed that engineering schools needed more government support, particularly 'mechanical appliances, testing machines and workshops'. Dillon identified TCD as one of the best schools in the Kingdom but observed that Irish engineering education lagged behind other countries. He stated that it was the duty of the engineers in Ireland, and of the Institution, to highlight the needs of the country.

Edward Glover's presidential address in 1900 comprised a detailed, and very useful, explanation of the 1899 Local Government Act. This act had replaced the older system of Grand Jury, Poor Law Board, Board of Works, local magistrates, Lord Lieutenant and Sheriff with County and Urban District Councils.

In the previous administrative system the Grand Jury had charge of County expenditure. The Jury was selected by the Sheriff of the County, who in turn was appointed by the Lord Lieutenant. Once discharged by the Judge of Assize, the Grand Jury ceased to exist. Presentment sessions were held in Baronies prior to the Assizes to consider works that would be ratified or rejected by the Grand Jury at the ensuing Assizes. There were also county-at-large Presentment sessions.

The Grand Jury dealt with: making and repair of roads and bridges; the construction and maintenance or courthouses; the county printing and the administration of a number of statutes.

The Grand Jury levied rates to support lunatic asylums, county infirmaries, industrial schools, coroners, certain constabulary charges, the conveyance of prisoners, guarantees to railways or tramways and other less important matters down to the payment of court criers and tipstaffs (court officials).

The Grand Jury also considered claims for malicious damage to property and other issues. The only check on the presentments by the Grand Jury was the Judge of Assize.

The Local Government Board & Boards of Guardians under the Poor Law system were: public sewer-makers, custodians of burial grounds and wells, the constructors of waterworks, the proprietors of dwellings for labourers, executors of compulsory vaccination laws, and the laws relating to the sanitation of dwellings and public nuisance, muzzling of dogs and slaughtering of diseased animals. The Poor Law boards levied separate rates from those of the Grand Jury. Under the new system all the fiscal business of the Grand Jury was transferred to the Councils.

Urban and Rural District Councils became the sanitary authorities with powers relating to: sewers, water supply, public lighting, labourers' cottages etc. They would employ engineers paid by fee, or sometimes salaried, and the qualifications of such engineers were beginning to become an issue.

The position of county surveyors had been established in 1834. If there was work for which no suitable contractor appeared, then such work could be placed in the charge of the county surveyor. By 1900 the county surveyor exams were no longer competitive, which Glover considered to be a retrograde step. Glover noted that the grade of Associate Member of the Institution of Civil Engineers of Ireland (AMICEI) was recognised as a suitable qualification for the post of Assistant Surveyor (without the need to sit an examination). Similarly, the grade of MICEI was accepted for the post of Urban Council – Surveyor: Glover credited this to lobbying by the Council of the ICEI.

He noted that Irish agriculture was behind the times, that flooding was a continuing problem that called for arterial and river drainage, and that more trees were needed. He also expressed the need for a Railway Department to ensure cheap transportation of goods. Noting that the P&T Department was state run, he considered it to have been a success and that the winning of peat was awaiting technological developments.

John Henry Ryan, who had been in college with John Purser Griffith, began his address in 1902 with a brief synopsis of the Institution's history. He called on all members to cease appending CE to their names, recommending instead the use of MInstCEI., AssocMInstCEI. or AssociateInstCEI. as appropriate. He recommended that the Institution should urge the Local Government Board to appoint ICEI qualified candidates, 'irrespective of political or religious opinions' on the basis that it was 'in the interests of the health of the community'. He the described the genesis of Civil Engineering, the separation of Mechanical Engineering and the subsequent divergence/emergence of new disciplines, identifying the common bond of all disciplines.

Ryan then presented a detailed list of the many recent key international construction projects around the world. He also discussed some current national rail projects and stated the need for increased government financial support. The then current requirement to have an existing railway company involved in any new railway schemes was cited as a problem because it was seen as a restrictive clause. As previous presidents had done, the generation of electricity from the harnessing of the river Shannon at Ardnacrusha and its distribution (what became known as the Shannon Scheme), which had just been approved by Parliament, was one of the Irish projects mentioned by Ryan. However, he also commented that there was no sign yet of the capital being made available for the project.

He advocated the state purchase of the existing inland navigation system to be followed by the state purchase of the railway system. He also suggested an extension of the Grand Canal scheme from Athlone to Galway. Like some of his recent predecessors he highlighted the need for arterial drainage work in Ireland and commented on the shortcomings of the private bill legislation that required that such bills be dealt with in London rather than Ireland.

In proposing a vote of thanks to the president for his address Berkeley Deane Wise wished that the ICEI could be seen in Belfast, 'if only once in a lifetime'. The Institution was at the time largely Dublin based and Wise thought that it would do a great deal of good if the Institution were to hold a summer meeting in Belfast and concluded by saying that 'it would bring prominently before the professional gentlemen in the North the fact that there is such an important and flourishing Institution'.

Beginning his presidential address in 1904, **Robert Cochrane** mentioned that an index to the Transactions had been completed the previous year and that the Institution's By-Laws had been changed to bring them into line with those of the ICE.

ThCochrane then described the Institution's visit to Belfast, a first in its history. Belfast City Hall was then under construction and they had visited the works. They went to Harland & Wolff and saw a new graving dock under construction. They took the Harbour Commissioner's steamer *Musgrave* down the river quays, and visited the Sirocco Engineering Works. They travelled by train to the Giants' Causeway and were fêted by the Lord Mayor. The visit lasted three days.

Cochrane then gave an account of the Society of Civil Engineers (founded 1771). He noted that this group saw 1760 as the start of 'a new era in all the arts and sciences, learned and polite, commenced in the country [Great Britain]'. This body was by then entirely social with membership limited to fifty individuals. The president also mentioned other engineering societies including The Engineering and Scientific Association of Ireland, which had been established in 1903. The President also gave an account of Cooper's Hill Engineering College. This college had been established in 1871 to provide engineers for the Indian Public Works Service, but was now costing more than university courses and so the college was being closed down. The presidential address has an appendix with considerable detail on different university courses and their costs.

Giant's Causeway Tramway near Dunluce Castle

He discussed the centenary of the railway locomotive and then gave a detailed account of the engineering works visited in Belfast, the Belfast water supply, Queens College Belfast, the Municipal Technical Institute, and the first hydro-powered electric tramway. The tramway ran eight miles from Portrush Station to the Giants' Causeway. It had originally been a third rail system until a cyclist fell on the line and was electrocuted which led to an overhead trolley system being substituted.

It appears that the next president, **William Ross**, received very little warning of giving his presidential address, hence his address in 1906 was short. He had in fact agreed to speak as there had been no paper scheduled for the evening in question.

Ross mentioned that the new Bye Laws contained a provision for the appointment of a paid official. There was a sense that the Institution was considering the introduction of an entrance exam and there was 'a good deal of discussion at present as to the State making provision for the registering of engineers'.

Cochrane, in proposing a vote of thanks believed 'that with such a distinguished president, and such energetic and efficient members, that they had before them a year of progress, which would surely equal and surpass any of the past years' workings'.

Ross had also suggested that branches of the Institution be formed outside of Dublin and James Dillon, in supporting the vote of thanks, said that 'undoubtedly the power of the Institution would in some degree depend upon the members' and that efforts should be made 'to bring in men who resided in the outlying districts of Ireland'. 'By doing so they would give greater publicity to the Institution, and circulate in the rural parts of Ireland the advantages that could be derived from being members of it'.

As to the then bare walls of the Hall at 35 Dawson Street, Berkeley Wise had some practical suggestions to make. He suggested that portraits of past-presidents be hung on the walls, with tablets giving their names and the years that they served. Wise said that he did not like the carpet on the floor, that the lighting could be improved, and that the sound was so bad that he had not heard a single word of last month's paper!

Joseph Henry Moore's address in 1907 started with a brief review of membership numbers, which were up, possibly because of the Local Government Board's decision to recognise Members and Associate members of the ICEI as suitably qualified persons for certain engineering positions.

The topic of Moore's address was "a brief sketch of the progress of Civil Engineering in Ireland and the prospects of employment for Civil engineers at present". He began with a history of road development in Ireland and presented many interesting facts including comments on regulations relating to carts with narrow tyres. He mentioned canals in passing before returning to roads and the bodies responsible for their construction and repair. He suggested that all roads should be under state control and mentioned Cox's bridge in Waterford and the compensation paid when it was to be reconstructed and made free to the public.

The president next discussed railways with an emphasis on their financing, although he also addressed electric traction. He mentioned a proposal to extend the existing light railway to the Arigna coal mines in county Leitrim. However, he noted that the county council chamber had been invaded by stick-armed carters who forced the scheme to be abandoned. Moore didn't have had a high regard for local councilors, but he said he had little time for mob law. Later in the address he observed that ...'to carry out public works the first requisite is money, and to get money there must be credit and credit is easily damaged by such proceedings (mob law etc.). So far however, Irish credit is good'.

He expressed his scepticism of a proposal to dock transatlantic ships in Blacksod Bay and transport passengers across the country by rail. This proposal seems to have been a variation on the nineteenth-century plans to dock transatlantic passenger ships on Ireland's west coast (such as Galway) with a rail link across Ireland to link with a ferry to the UK.

He believed that canals and drainage were for the most part incompatible, identified drainage as more important, and discussed arterial drainage in some detail. He praised the development of Dublin Port, but did not think further money should be spent on the smaller east coast harbours, given the low level of traffic. He noted that models of the present and proposed [Dublin] docks were to be seen at the Irish International Exhibition.

He then discussed the state of water supply in Ireland. It seems that at the time most towns already had, or were installing, water supply systems. He made particular mention of Trim and its Hennebique RC tank, commenting that this was the third such structure of its type in Ireland. However, many small villages did not have a public water supply. He also discussed sewage treatment, being sceptical, even of the need for treatment in the case of most of the smaller towns and villages where sewage

Reinforced concrete water tower at Trim 1909   could be discharged together with rain runoff into fast-flowing water courses. He described Dublin's main drainage as a success but saw the primary settling tanks at Pigeon House as problematic due to the release of odour. Belfast, he noted, discharged sewage into Belfast Lough and Cork was currently without intercepting sewers.

George Murray Ross reported in 1909 on the continued increase in the Institution's membership, noting that the number of Associates had dropped slightly. Mr. Ross thought this was as it should be, the grade of Associate being described as of "more elastic definition." Ross urged members to be conscientious in signing application forms, indicating that members should ask candidates to state what grade of membership they are seeking – it should not be the task of the Council to have to investigate. Associate membership could now be obtained through examination. Ross noted that the syllabus for these examinations recognised that the definition of a Civil Engineer in the Institution's bye-laws included civil, mechanical and electrical engineering. The examinations were not intended to take the place of a regular engineering education, but to provide the Council with additional information.

Ross next commented on the best education for an engineer. He advised against early specialisation arguing that an engineer needed basic competence across a range of fields, besides specialising early could restrict choice in one's later career. He continued:

> 'A good engineering education, from a professional point of view, has two functions: the giving of information and the teaching of how to use the stores of knowledge already accumulated, and if possible increase them. The attempt to supply detailed information or information on all the numerous branches of engineering only leads the student to mistake details for principles'.

He saw the fundamental sciences as key, with mathematics as absolutely essential. However, he also valued working knowledge and believed that students should serve an apprenticeship after their college course. However, he also saw college as 'the place for learning the art of measurement, which is the true subject of laboratory teaching'. He quoted Bindon Blood Stoney who considered "exactness" the key quality most essential in a young engineer. He didn't think that 1909 was a great time for young engineers in Ireland, given the lack of prospect of further railways and canals being constructed, but he saw prospects in the wider world as encouraging.

St John's Bridge, Kilkenny (Reinforced concrete 1912)

The remainder of Ross's address looked at how engineering had advanced in the previous forty years. He considered how the quality of steel had improved and the rise of the internal combustion engine, the automobile and the aeroplane. His observations on the then empirical nature of aeroplane design make for interesting reading. He went on to discuss a range of issues from road design to the advances in electric lighting. He commented on sewage treatment and included some detailed and interesting remarks on septic tanks.

Ross discussed ferro-concrete, but was clearly wary of a material where a structure 'may collapse owing entirely to careless workmanship'. He mentioned the new reinforced concrete bridge in Kilkenny and proposed plans for Waterford (when it was completed in 1912 the new bridge in Kilkenny was the longest single-span reinforced concrete road bridge in the British Isles).
Throughout his presidential address Ross identified potential sources of employment for young engineers, these ranging from general industry to the construction of labourers' cottages.

At the start of his presidential address in 1911 **Peter Chalmers Cowan** quoted Sir Benjamin Baker, who questioned the continuing need for the traditional presidential address. In his opinion, in the early days of the profession, when technical literature was non-existent, the addresses constituted a valuable medium for disseminating useful knowledge and for that reason were not only listened to, but read by engineers, but at the end of the nineteenth century the conditions were vastly different.

However, in the event he gave a traditional lecture and was thanked for doing so by past-presidents John Purser Griffith and Robert Cochrane.

Cowan's address continued the discussion on the desirable attributes of an engineer, which was started in the previous president's address. Cowan distinguished between engineers and craftsmen and discussed the education of an engineer at some length. He deplored poor spelling and stated that accurate expression was of the greatest value. He was in favour of laboratory work to help train engineers but was not in favour of extensive manual training because it was not feasible to train an engineer in all the professions and not all workshops would give students a broad overview. Cowan noted that modern graduates may now get paid a modest amount during their time as a junior assistant. This was a new development. Previously junior engineers often paid for their training.

Cowan believed that the courses for engineers in Irish Universities were generally excellent, but that the pass standard was rather low. University education was still treated with scepticism in some quarters, but the recent Conference on the Education and Training of Engineers, organised by the ICE, confirmed the president's view that a university degree, though not essential, is of great value to an engineer. The president drew a distinction between passing an examination and pursuing a college course.

The need to restrict engineering to qualified persons was stressed. It was Cowan's belief that engineers needed similar protection to that of lawyers and medics. He held that public money and public posts should be administered and held by qualified persons: 'There lingers in many influential quarters the idea that the construction and maintenance of roads, and the provision of water supplies, and sewerage schemes can be placed in the hands of uneducated, untrained, and unskilled men'. He highlighted the contrast between British and German attitudes to technical experts.

The president then moved on to discuss the difficulty of predicting the future, for example, Robert Stephenson's doubts over the feasibility and commercial value of the Suez Canal. He remarked that it had been said that "the future does not come towards us, but streams from behind over our heads", and he continued: 'anyone who studies the history of engineering will be slow to set a limit on the progress of our profession'. Cowan ventured to suggest that the best remedy for pessimism was the study of history, 'from which it will appear that all the present and threatened evils which many of us think are unprecedented and likely to overwhelm us, were experienced and surmounted by previous generations in many lands'.

Cowan went on to mention a number of previous papers published in the Transactions and discussed possible progress on such diverse topics as burning peat for fuel, road making in bogs, train safety, a proposed steamer for "sweeping" the bed of the Liffey, main drainage, coastal erosion, water supply works, advances in water sterilization using ultra-violet rays from electric lamps, the new bridges in Kilkenny and Waterford, ship building in Ireland, the internal combustion engine and Charles Parsons' development of the steam turbine.

Cowan also discussed the Parliamentary Report on Canals and Inland Navigation, which recommended the retention and improvement of the canals.

Surprisingly, traffic on the canals had increased between 1888 and 1905. He felt that the state should acquire the canals under a central board of control. A vice-regal commission on railways had recommended that the railways should be nationalised.

William Collen, delivering his presidential address on the 5 November 1913, started by saying that the members would have an opportunity to remove him or re-elect him next May. The relevance of this remark was that up until then, in accordance with the Institution's by-laws a president might hold office for two years and that had been the practice since 1859.

The number of members of all classes was 352 but likely to remain below 400. However, there were still sufficient numbers to rotate the presidency on an annual basis. The president suggested that this idea was not new, but could only be raised by the president himself without potentially seeming to cause offence.

33

Collen's address had two themes: a special look at road construction, specifically a report on the Third International Road Congress held in London; and secondly, an overview of past projects and a number then under construction, including the Panama Canal, which in 2014 celebrates the centenary of its opening. He then described the Road Congress by detailing the topic areas and the papers delivered. He advised that a complete set of the volumes containing the Minutes of the Congress had been placed in the library of the Institution.

He proceeded to review some recent advances in engineering. He first mentioned the manufacture of patent fuel and that marine engineering steam turbines were gradually displacing reciprocating engines. He noted that the use of oil fuel, especially in warships, was increasing. It seemed at that time, even according to Charles Parsons, that until oil prices were reduced below 23 shillings per ton, steam raised from coal was cheaper. He remarked that during the past ten years the great increase in motor traffic had made the introduction of some method of preventing dust nuisance an absolute necessity.

The president mentioned gas works and the "liming" of coal before carbonisation. Despite the almost universal use of electric light, the number of gas consumers was still increasing.
Some UK railway companies had electric working on some short sections of their systems. Direct 3,500 volt current was used by the LSWR (Waterloo and Guildford and lines serving Brentford, Hounslow, Twickenham, Shepperton, Kingston and Hampton Court.)

Collen considered that the most striking advancement in land-based telegraphs was the increased adoption of the Baudot multiple-telegraph system. This system allowed as many as eight messages to be sent simultaneously down one line. There had been no improvements in the public telegraph since the Postmaster General took over on the 1 January 1912. Marconi was mentioned as the front runner in this new technology but the Post Office was 'not quite happy' having recommended the Marconi Wireless Telegraph Company for connecting the Empire, as there were alternative systems becoming available.

On the topic of aeroplanes, Collen noted that Adolphe Pégoud had flown upside-down, commenting that 'the limit of command over these machines appears to have been reached'. He mentioned Harold Hawker's attempted 1,600-mile circuit of the UK over water, which was unsuccessful due to forced descent at Loughshinny. He also remarked on the appalling disaster that had befallen the Zepplin IV on her trial trip the previous month.

Whilst noting that Belfast continued to sustain her world-wide reputation for shipbuilding, Cowan mentioned other industries in the area producing such items as pumping plant, tea-drying apparatus, and ventilating systems.

He then referred to the 'vast powers' contained in Ireland's bogs and rivers and reckoned that one square mile of ten-foot deep bog was the equivalent in heating power to 300,000 tons of coal. He also commented on the hydroelectric potential of the river Shannon and smaller rivers with good falls, but cited fisheries, navigation, and riparian owners as barriers to the Shannon scheme.

Collen drew attention to the cutting down of trees and identified the sale of land to tenants as the origin of the problem. He continued:

> 'This vandalism is to be deplored, not only on aesthetic grounds, but on the higher grounds of the nation's temperament, which is closely connected with its climate. For forests tend to moderate the extremes of climate and to increase the rainfall, and their destruction is invariably followed by a diminution of the rainfall'.

The president noted the findings of the Royal Commission on Canals and Waterways that, in 1906, recommended a central waterways board. Past-president, John Purser Griffith, in a paper to the British Association, had advocated the revival of inland navigation with the aid of public funds, but there had been opposition from the railway companies to the suggestion that low-class freight could be moved from the railways to the canals.

He identified the Channel Tunnel as being the most important engineering work likely to be undertaken in the near future, but cited military fears as the bar to progress. He then gave a brief history of the various plans for a tunnel, the final plan he described comprising a pair of single-track circular tunnels with cross-passages (not dissimilar to the tunnel as opened over 80 years later in 1994).

At the time of the president's address, the general strike and lockout in Dublin had been in progress for three months and he referred to the capital as 'a city of the dead'. He was critical of the unions, but saw the strike against a backdrop of 'more or less continuous conflict between capital and labour'. He believed that some form of co-operation or unity of interest was desirable and went on to suggest the establishment of 'joint trade committees' which, from his description, sounded very much like the present Labour Court. He continued: '...employers, from behind closed gates, look out at their departing industries – outside men stand about, aimlessly, sullen, in enforced idleness, while above the din of battle are heard the cries of starving women and children'.

Mark Ruddle began his address in 1915 by noting that he was the first member of the "Electrical Branch of the Engineering Profession" to occupy the presidency. The address was the first delivered during WW1 and the war dominated the early and middle sections of his address. Ruddle saw the war as being a war of engineers on both sides and he made very interesting and insightful comments on the industrial and commercial backdrop to the war. He saw German industry and industrialism as a model of efficiency, but suggested that they had been pursuing a policy of....expansionism. One may interpret Ruddle's comments as indicating that Germany was selling products abroad at reduced cost to capture greater market share. He already seems be imagining a post-war period when the cost of the conflict must be borne by society. He wonders, and worries, about the capacity of the allies to supply their industrial needs without a dependence on Germany. Germany was at the time clearly the enemy, but there was a real sense of Germany as being a threat to the free world and civilisation – in contrast the president saw the British Empire as having come into existence as a natural outcome of the spread of the principles of civil liberty.

It is well worth quoting some of Ruddle's remarks 'in extensio'. He asked first, what his audience considered to be the real issue being fought out on the fields of Belgium and the plains of Poland and continued:

> 'It is not merely a material one; it is a conflict between human ideals, the result of which may determine for centuries to come the moral progress or retrogression of our race'. 'This flood of death and destruction, sweeping away with it the hallowed sanctuaries and garnered treasures of a continent, annihilating the spirit of its past, blasting the promise of its future, and bringing ruin and life-long tears to millions of our fellow-men, is to determine whether force of freedom is to govern the future of our race, whether brute force is to be king; or whether the right of humanity to follow its ideals and the dictates of the moral law is to be supreme'.

The president considered war production and the likely post-war world in considerable detail. In particular, he considered the problem of the conflict between capital and labour. He took a balanced view, seeing both as controlled by market values. However, he saw the justice in a living wage that must include 'in any case, food, shelter, fuel and clothing as absolute necessaries, but it should include some provision for the enjoyment of life according to individual ideas or tastes'.

Ruddle then said a few words in favour of vocational training and held up the technical training given to navy personnel as a model. He discussed the proper course of educational training for an engineer and cited a recently issued report of the special committee of the London Institution (ICE), which he said showed that there was still a wide difference of opinion amongst leading engineers on the subject, comments that still ring true. He said that modern graduates were not good at tackling new problems, but did not see a necessity for a modern engineer to be a trained mechanic. He saw benefits in a firm general theoretical base and felt that a student with such a previous grounding would thus gain more from workshop training.

Walter Lilly began his address in 1917 by noting that he was the first president who simultaneously occupied an official position in Trinity College Dublin. He mentioned that in order to help the war effort he had for the present to reside in London. 'No war hitherto has been dominated to such an extent by the production of the engineer'. Lilly said that it would be an impossible task to review the production of munitions during the war. Therefore, in an effort to try to forget the war, he discussed the development of the rotary engine. Later in his address he stated that this choice was also influenced by the fact that he had to put it together in his spare moments whilst undertaking munitions work.

TCD Engineering Laboratories 1903

Lilly gave a brief historical outline of the development of a rotary engine that he had built in the autumn of 1914 in the Engineering Laboratory in TCD. His engine used a more perfect form of gearing, amongst other features. His results having seemed promising, he had taken out a British patent in 1916. Unfortunately, he found that his gearing had been described elsewhere, so he abandoned his plans to take out foreign patents. Lilly argued for a "patenting" system that could protect designs. He highlighted the problem of having a successful design copied by imitators. He said that patent agents of experience supported this view. The Patent and Design Law gave protection at one end of the scale and Design Registration protected 'the mere external appearance of, say, a piece of engineering apparatus'. However, these regulations left design in the middle and largely unprotected.

Lilly went on to consider the three types of motor: reciprocating engines; steam turbines; and rotary engines. He then touched on the relationship of research work to engineering and classified engineering research under three headings: research undertaken purely for the advance of engineering science; research undertaken because there is money in it; and research to determine physical constants of materials, standardisation of gauges, screws (eg National Physical Laboratory).

Lilly, in common with Ruddle, forsaw 'the industrial war that is coming in the future'. On inventions, he called for a department to 'offer assistance to the inventor or in helping to bring the invention before interested parties who would undertake its development'.

Lilly also noted that the war had shown that '... the most delicate and accurate mechanical work is well within women's capabilities'. He concluded by listing the technical advances that had resulted from the war.

The next president, **John Ousley Bonsall Moynan** noted that he was the first president to be chosen from the body of country members (those residing 100 or more miles from Dublin - he lived in Nenagh). Speaking in 1918, Moynan noted that is seemed probable that peace would soon be declared (three days later the Treaty of Versailles ended WW1). He continued: 'Many of our members, young and old, have taken service in His Majesty's Forces, and have rendered, as was expected, a good account of themselves. Some have been wounded, and some alas! have given their lives in defence of their country,...'.

The topic of the president's address was inter-communication in Ireland and Moynan considered the different modes separately. Firstly he gave a history of road and road-based transport development in Ireland. Next, he considered canals, again tracing their history, before moving on to railways.

Moynan then analysed the current situation. He saw no future for canal transportation, or at least no future in terms of further expansion of the canal network. With respect to railways he noted that some people believed in the potential expansion of the network and they pointed to the fact that some districts are almost sixty miles from the nearest railway station. However, the cost of construction, under 'war-time charges', was approximately double the normal costs. In addition, the problem of mixed gauges resulted in increased handling costs and increased damage. Moynan thought that railway expansion was unlikely. He saw the internal combustion engine as the solution and recounted the famous couplet:
"Canals were, Railways are, Roads will be
The most important of the three".

Moynan proposed establishing the equivalent of the current NRA – an "Executive Road Board" – with responsibility for all principal roads. He highlighted the benefits of lower costs, flexibility and door-to-door service available to road transport. He suggested that: 'if you have a heavy class of car, you must limit the maximum speed to about eight miles per hour, otherwise cars will cut up any roads you make. Lighter cars, intended for the carriage of mails and passengers, would be allowed a speed of fifteen miles per hour'. He concluded by opining that the future hope of increased inter-communication in Ireland will depend on the motorcar and the internal combustion engine.

Arterial drainage

The next person to occupy the presidency, **Patrick Harnett McCarthy**, began by drawing attention to Sir John Purser Griffith's recent election as President of the Institution of Civil Engineers. Speaking in 1919, McCarthy mentioned the part played by engineering skills during WW1 but noted that after five years of zero development, the needs of the country had to be addressed, but against a backdrop of 'scarcity of materials, ... enormous taxation, and ... high cost of labour'.

He announced that the Institution of Civil Engineers (ICE) had a Bill under consideration for the registration of civil engineers, which was to be presented to Parliament and that the ICEI was to cooperate with them to secure its enactment. He felt that, as doctors and lawyers were registered in the public interest, so too should engineers. 'It is a recognised fact amongst qualified Engineers that only those whose professional qualifications are open to doubt make use of the letters C.E.'. He suggested that the Local Government Board 'should not place public works under such men and judges should not take evidence from them as engineers'.

The President next spoke about the need to change the method of training engineers. He stressed the need for a firm scientific training. He stated that most young engineers now attend engineering schools attached to universities and following three years of study and the passing of exams obtain a degree. He urged that 'the young engineer leaving college

should endeavour to get on to practical work at once'. He believed that the conferring of the degree should be postponed until after the student had gained experience and pointed out that this was the case with medical students. As a result of the war, the outlook of boys is very different from pre-war days. 'The doer will be the man of the future, not the dreamer.'

Having discussed the disposal of sewage at some length, McCarthy proceeded to discuss water supplies and noted that there was 'insufficient storage, lack of elevation of reservoirs, small mains, shallow wells close to badly constructed sewers (and that) owing to the more general use of baths and the substitution of water closets for the old privy, ample provision should be made in the size of the mains to meet these contingencies'. 'Water engineers in this country are badly handicapped for want of sufficient rainfall records'. The Institution had made a suggestion to government that rain gauges should be distributed to police stations, where readings could be taken and the data readily made available. He also suggested that stream gauges should be installed to get information useful for water supply and water power. He referred to a recent paper by Chaloner Smith on the flow of the Shannon in which the daily flow over many years was presented, data which became invaluable to those who in the 1920s were charged with the design of the Shannon Scheme. McCarthy hoped that the government committee on Water Power in Ireland under the chairmanship of Sir John Griffith would take steps to gather such flow information on other rivers. The benefits of hydropower were acknowledged but the president advocated a rational approach where the costs justify development. In such circumstances fishing rights etc. should be dealt with equitably. He also remarked that the increased use of agricultural machinery should promote small repair and manufacturing workshops.

On arterial drainage, McCarthy mentioned that large sums had been spent on the survey of the Barrow catchment and on preparing plans for its improvement. He went on to explain the complex issues involved in arterial drainage and noted that the improvement of the lands is not such as to justify the expense, hence state funding is necessary. He justified such state support on the basis of the security of food supply and made the somewhat unscientific prediction that 'if the water were drained away more quickly there would be less evaporation, less cloud, more sunshine, more heat radiated from the earth and a higher summer temperature' and 'there would be less pulmonary disease, less rheumatism, and less pre-disposition to disease generally'.

McCarthy next dealt with the advent of the motor car and noted that cars require tarred roads (not suitable for horses) and that is was necessary to preserve the road and to keep down dust. He noted that 'during the recent railway strike, as well as during the war, the heavy motor vehicle has proved its worth'. McCarthy ended by giving the total membership of all classes in the institution as 384 and encouraging members to persuade non-member practicing engineers to join.

Francis Bergin in 1920 began by noting that, in the early days of the profession, many engineers received their training as apprentices. However, in the age of steel, reinforced concrete and electricity, he felt that a university training, which gave a sound theoretical and scientific training, was essential.

Bergin mentioned that a Bill for the registration of Civil Engineers (London), prepared during Sir John Purser Griffith's presidency, was to make provision for the registration of all ICEI members. However, it was opposed by some English societies and associations, which were not treated so favourably. As a result it had to be withdrawn.

A past-president, J.O. Moynan, had been responsible for convening the Joint Council of Executive Professions to report on post-war reconstruction. The Council, meeting on 8 April 1919, comprised architects, county surveyors, electrical engineers, mechanical engineers and borough surveyors. The council identified the work that needed to be done, including provision of adequate water supplies and sewerage, road construction, utilisation of water power and peat, arterial drainage, developing coal and other mining, constructing harbours, railways, and strengthening bridges, and provision of houses for the working classes.

Bergin presented extracts from one of the Council's reports, but because no detailed schemes were ready for consideration, the reports were generalized. In essence, the council had agreed that Britain needed industrial reconstruction – Ireland needed industrial construction; the wide difference in the problems (Ireland and Britain) required widely different solutions; and ergo: Ireland should get a proportionate share (on a par with Great Britain) of the total sum and that this money should be administered by Irish business and professional men with experience of Irish conditions.

Bergin noted that no money had been spent so far on the schemes listed above by his predecessor McCarthy, and the prospects for funding were slim. The cost of providing housing exceeded the income obtained in rent. Thus housing constructed under the recent housing act would lead to rate increases. He felt that the housing act was also deficient because it made no provision for towns that were under the control of rural councils.

Next to housing came the need for improved sanitation. Bergin painted a picture of ineffective legislation and lack of powers to force owners of slum properties to improve the sanitary conditions of their tenants. He felt that all the Public Health Acts should be scrapped and replaced with a simple code prepared by a committee of practical men. There should, in addition to a county surveyor, be a county engineer and a county sanitary officer.

Joshua Hargrave, speaking in 1921, referred to the increased costs of commodities since WW1, in particular newsprint and the resulting cost of newspapers. He noted that the Institution's heaviest expenditure was on the Transactions, but that, nevertheless, it had been found possible to maintain their publication throughout the war. The question of registration of engineers was still very much on the agenda and Hargrave noted that the public were still inclined to regard the letters C.E. to have the same authority as M.D. John Purser Griffith, when president of the Institution of Civil Engineers had taken great interest in the matter of statutory protection of engineers and had proposed the use of the word "chartered".

Hargrave continued his address by alluding to the fact that there were currently some sixty different loading gauges in use in Europe and that this prevented continental wagons from running on British railways, a problem whenever a tunnel was to be constructed under the English Channel (although the unification of railway companies did much to eliminate the problem, basic residual differences had still to be resolved when the tunnel was eventually completed in 1994). Referring to Ireland, he felt that the narrow-gauge lines could have been built at standard gauge for little extra cost. Any extra cost would have been confined to three items: extra width of earthworks, extra ballast, and larger sleepers. He cited the case of the narrow gauge extension to the Arigna colleries, which had resulted in the need for double handling of coal being transferred to the standard gauge network. Hargrave suggested the use of home-produced concrete sleepers as an alternative to timber sleepers, but at the time they had not proved successful. It is only in more recent times that concrete sleepers of improved design have largely replaced timber. As train loads increased, there was renewed debate as to the relationship between live and dead loads on bridges, in particular the effect of impact loading. It was generally accepted that, for design purposes, live load should be taken to be about twice the dead load. Mechanical traction had come to stay and it was considered that many bridges were not capable of carrying the increased loads.

Hargrave suggested that it is only by looking back that one can come to realise the progress that had been made. One hundred years ago (1820s) there had been no steamers, no trains, no gas-lighting, no telegraph, and that, within the lifetime of many in the audience, have been invented and brought into use electric lighting, electric traction, the telephone, wireless telegraphy, the motor car and the aeroplane. He felt that it seemed almost impossible that similar progress could be made during the next century, but that in all probability it will be accomplished. He concluded by summising 'I can quite imagine our descendants looking back to this period and wondering how we existed without so many applicances which they will have come to regard as actual necessities'.

In 1922, when **Pierce Francis Purcell** delivered his presidential address, around one-third of members were graduates of the various engineering schools. During the period of WW1, the Rising and the Civil War, the work of the Institution was carried on under great difficulties, but it had been found possible to keep the Transactions going. Purcell referred to the first paper read by Robert Mallet on the 12 November 1844 on "the artificial preparation of turf, independently of season or weather" and noted that there was still no sign of a solution to the problem. Purcell had been a member of the Irish Peat Enquiry Committee in 1918 and later was to be involved with a burgeoning peat industry.

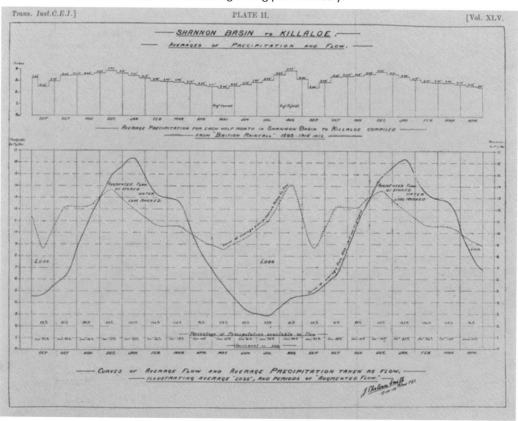

From Chaloner Smith's paper in the ICEI Transactions on Flows in the Shannon

Purcell also highlighted the importance of the Transactions as having recorded the work of Robert Manning in the development of a formula for the determination of the flow of water in open channels and pipes, an empirical formula still in use today for open channel flow calculations. Purcell noted that the research on the strength of columns and struts carried out by Dr Lilly at Trinity College Dublin with totally inadequate funds and equipment had established many points of fundamental importance.

In Volume 45 of the Transactions, Chaloner Smith had set down what proved to be invaluable data on the average volume of flow from large catchment areas in Ireland. The data was available to McLaughlin and Siemens Schukertwerke when planning the Shannon Scheme. Reference was also made to two important papers published in 1898 by Purser Griffith on "Portland Cement". These dealt in detail with the properties and testing of cement, and were considered to contain ideas far in advance of those then held in Britain. Purcell noted that, in the previous fifty years, there had been few civil engineering works of importance executed in Ireland outside of waterworks and sanitation. Noting that reinforced concrete design was mostly carried out in London, often employing patented systems of reinforcement, he saw no reason why the design could not be carried out in Ireland without being bound to any particular system.

On the matter of registration of engineers, the ICEI Council had set up a committee to consider the issues. Purcell felt that there should be no *grandfather* clause allowing persons to be registered who were presently holding office, but who were not properly qualified. He urged the employment of Irish engineers where possible, rather than relying on overseas consultants, and remarked that 'nothing can be more mischievous from the National standpoint, or more discouraging to the members of the profession, than the absolute handing over of works to outside firms of engineers'. On the vexed question of pupilage, he held the view that the best-qualified men from the engineering schools were worthy of their places in the engineer's office without fee and should be self-supporting.

At a time when the country had been entrusted to the people of Ireland, Purcell was naturally expected to devote some attention to the question of the increased development of its natural resources. Purcell perceived that the need of the moment was to get the country back to normal conditions and to provide work rather than unemployment and dole queues. Apart from the task of repairing roads and bridges damaged during the Civil War, there was a need to invest in arterial drainage and to develop the hydropower potential of the country and to utilize the vast reserves of peat fuel. It was reckoned that coal mined in Ireland amounted to around 100,000 tons per annum, as against some 4.5 million tons imported, representing some 55% of Ireland's heat requirements. It was estimated that deposits of peat had the potential to produce 4,000 million tons of air-dried peat. The 1918 report of the Irish Peat Engineering Committee and the experiments with the machine winning of peat, especially at Turraun, had laid the foundations for a prosperous peat industry, but it was not until 1934 that the Turf Development Board was set up, morphing into Bord na Mona in 1945.

As the senior vice-president, **Arthur Hassard** was, due to his official duties, unable to go forward for election it fell to **James Thomas Jackson** to take on the presidency in 1924. He began by acknowledging the contribution of his predecessor, Professor Purcell, who had agreed to carry on his valuable work as Honorary Treasurer of the Institution.

Having defined the role of the Institution and that of a Civil Engineer, Jackson went on to discuss the need for standardization of minimum qualifications through registration. He suggested that there was a need for a professional licensing body to be set up by the State. The principal duties of such a licensing body, he suggested, would be a) the keeping of a register of qualified members of the profession, b) assisting in the maintenance of educational standards at professional schools, c) acting as a disciplinary body for the profession, and d) determining the qualification necessary for registration or recognition. All these proposed duties were in time to become functions of the Institution and the use of the title "Chartered Engineer" became the method of signifying qualification as a professional engineer. Recognition of educational awards and their validation through accreditation ensured the maintenance of educational standards. A code of ethics was introduced to underpin the Institution as a disciplinary body, and membership rules and regulations determined the qualifications necessary for membership of the Institution, rather than State registration.

Jackson defined engineering as a branch of science, and the engineer as a scientific man just as much as the chemist, physicist, geologist or botanist. He quoted George Francis Fitzgerald (1851-1901) as describing engineering as being 'simply physics on a large scale'. Jackson felt that this was true, but that the aims of the engineer and the physicist were different; 'the physicist aspires to know, while the engineer desires to do; the physicist is rewarded if the results of his work are new, the engineer if they are useful'.

Jackson regretted the proliferation of professional associations and societies and the general tendency towards separation rather than amalgamation, towards division rather than union. He cited societies representing Civil, Mechanical, Electrical, Mining, Marine, Automobile, and Aeronautical Engineering; societies of Naval Architects, Locomotive, Railway, Signal, Colliery, Municipal & County, Sanitary, Heating & Ventilating, and Illuminating Engineers. He continued by noting associations of Consulting, Consulting Marine, Production, Inspection, and Structural Engineers, and many other minor groupings. He concluded that 'cooperation, if not federation or union, is certainly needed in the engineering profession'. Recent decades have seen a number of such unions and the demise of many of the smaller specialized groups.

He had some advice for the parents of candidate-engineers as to the suitability of a boy for the profession:

> 'There is a need for a careful and judicious estimate of the boy's inclination and suitability before deciding on the profession'. 'A lad who aspires to become an engineer should have a good general education, be "not bad" at mathematics and possess a sound constitution'. 'He should be active, energetic, not afraid of hard work, and willing to undertake a dirty or unpleasant job, for engineering is not a ladylike profession'. 'He should have a capacity for teamwork (usually exhibited in games), should get on well with his companions, should possess good common sense, and the natural ability to take advantage of opportunity'.

He felt that

> 'common sense was more important than scholastic brilliance'. 'The successful engineer usually has more difficulties with men than with materials, the energies that he directs are more often human than mechanical'.

In closing, Jackson noted that, whilst the engineer has 'applied the great sources of power in nature to the use and convenience of man', he is, in so doing, using up and exhausting these great sources of power.

> 'We of the modern era of progress have raided nature's storehouse of energy, and are now prospering on the proceeds'. Environmental concerns have grown over the ensuing years, but we are still 'raiding nature's storehouse'.

Arthur Hassard, speaking in 1926, began by noting that the total membership was 393 (266 in Ireland and 127, approximately 1/3 abroad), but that there were still many qualified engineers who were not members. Qualification requirements had recently become more stringent and he pointed out that the Council relies on the statements of proposers and if a member has doubts, whether he signs or not, he should acquaint the Honorary Secretary.

Hassard championed the development of specialism in engineering, likening engineering to medicine. He saw the reticence of general engineers to call in specialist colleagues as a reason why some clients go directly to manufacturers or suppliers. He felt that this may not be in the client's best interests.

The president said that he would have liked to speak on the construction, maintenance and management of the smaller harbours, but because a government tribunal was currently examining conditions at some of these harbours, and given that his experience in connection with Irish harbours has been under a government department, it would be highly improper. Instead he spoke of public works in foreign countries. This comprised a description of the organisation and funding arrangements in Norway, Sweden, The Netherlands, Belgium, France, Germany, the USA and Argentina.

Alfred Delap chose "Engineering in Ireland" as the subject of his presidential address in 1927 and took a look at the past, the present and the future. Towards the start of his address he commented

> 'We find ourselves at the moment at a very critical point in the history of the country and consequently, of our Institution. The political storms of the recent past left an ugly sea running for some time after they had died away, and a pitiful record of damage and destruction done – but most of the damage and destruction has been made good, the political sea is going down fast, and we all hope we are in for a long spell of quiet weather and a steady and prosperous voyage into the future'.

Delap began his discussion of the past by remarking on the youth of engineering as a profession. He recognised many early mistakes, both technical and administrative, but was reminded that 'we learn from our mistakes'. He noted that Ireland had no Roman period, but acknowledged the effort that must have been involved in the construction of raths. He noted that communication in Ireland had been discussed in the Irish parliament in 1700, but nothing had been done. In 1708 the Ballast Office was established to improve the administration of Dublin Port. (Britain was not much more advanced at this stage. The steam engine had been developed, but was not as yet widely used and, with the exception of water and windmills, there was little or no machinery). Delap identified the history of Dublin Port as the best illustration of the progress of engineering in Ireland. Prior to the first works in 1711, the port was 'a wide shallow bay blocked by two sand banks, the north and south bulls with the Tolka (river) on the north side and (river) Dodder on the south forming great muddy flats behind them...no shelter in storms and not sufficient depth'.

Although a channel was defined in 1711, the first substantial effort at entraining the channel was in 1748, when two stone walls with rubble infill were constructed along the south side of the river channel for a distance of approximately 8,000ft as far as the Pigeon House. In 1761 the South Bull Wall was begun by building the lighthouse at its eastern end, then working back to the earlier work at Pigeon House. This work comprised two walls of squared granite with rubble infill and a heavy granite slab surfacing built in deep water on sand, mud and shingle. The lighthouse was finished in 1768 and the wall in 1796. In 1819 construction of the North Bull Wall was commenced to the plans of Halpin and Giles. It comprised an embankment of massive granite boulders dumped at random. In 1822 Thomas Telford was consulted and found that the works were successfully increasing the depth of water over the bar. As a result Halpin's suggestion that the wall be extended was adopted and the whole work was finished by 1823. The first steam dredger was put to work in 1832 to maintain the navigable channel approach to the port.

On the assumption that Dublin would never be more than a tidal port, John Rennie was commissioned to construct an asylum harbour at Dun Laoghaire to the south of Dublin. When Rennie died in 1821 his son, later Sir John Rennie, took over the project. The Navigation Board (1730-1787) was criticized by Delap because none of the 22 of 23 different schemes it started was ever completed. He mentioned that the Newry Canal, begun in 1732, was the first summit-level canal in these islands. The canal between Newry and the Upper Bann was opened in 1741. He then listed a number of other canal schemes, including the Lagan Navigation (1753-1843), the Nore Navigation (1755-64), the Barrow Navigation (1759-), the Grand Canal and branches (1755-) and the Royal canal (1789-). The president saw both canals as a 'lamentable page of our Engineering History', and by 1859, even Mullins was describing them as unnecessary.

Delap next dealt with Ireland's roads. (also dealt with in some detail by previous presidents). Following a brief look at the development of Ireland's railway network, he moved on to the situation pertaining at the time of his address (1927). The Shannon Scheme was the only major scheme then completed, but water supply and sewerage works were being undertaken and the arterial drainage of the Barrow catchment was under way. Delap noted that there were a great deal of road projects and that communications had developed rapidly, but said 'who knows what is to come'. Delap predicted that 'while long-distance passenger traffic and that between big towns, heavy goods of all descriptions, including coal, will still be carried by the railways: other goods will be taken by road'.

He then discussed the Shannon Scheme. Interestingly his concern was with using the electricity generated by the scheme. He foresaw a problem that, if the cost of electricity was kept low to attract users there would be a deficit, but if the price was not kept low usage would be low and the scheme would be deemed to be a failure. He opined that 'unless the use of electricity tends to increase production, or to reduce imports, it will be of very little value to the country'.

Shannon Hydroelectric Power Station at Ardnacrusha, county Clare (ESB Archives)

Turning to the future, he made a number of predictions, most of which have come about. He predicted that 'our railways are not likely to be extended and non-paying branch lines will be abandoned; roads will improve and traffic densities will increase; no hope of a canal revival; trains will be replaced by buses; shipping will not increase in size'.

Delap hoped to see one big development – the construction of a transatlantic port on the west coast. He saw the port dealing with mail and passengers, but a past-president had dismissed the notion.

He saw the benefits to shipping of wireless communication and predicted that air traffic would increase requiring more landing facilities (he saw sea planes or amphibian types as most probable). He saw some challenges ahead, such as durable road construction, cement that was safe to use in seawater, and the development of our bogs as a resource. He ended his address by pointing out the benefits of the Institution's library, but noted that it was not much used.

The incoming president in 1929, **Michael Buckley**, acknowledged that it had been the custom to open presidential addresses with a reference to the history and progress of the institution, so he noted the rise in membership to a total of 389 members. However, in the census of April 1926, 1,325 persons (including two ladies!) had listed themselves under various engineering and naval architecture disciplines. He suggested that was a need to increase the proportion of the profession who became members of the Institution, noting that increased numbers would benefit all and would make it easier to regulate the profession and limit it to those who had received a recognised course of professional engineering education.

He was especially anxious to attract young members and
> 'here I should like to explode a crusted old diehard fallacy, that papers read before a body like ours ought only to contain descriptions of important works that are being, or have actually been, carried out. Such papers are excellent; but let us have, in addition, ideas, suggestions, proposals, plans, specifications, and estimates, if you will, of useful schemes which exist in the fertile minds of our gifted young members'.

Buckley extolled the benefits of electricity and predicted that domestic electricity would eventually be as standard as domestic water supply. Other forms of power were competitive in particular fields, but electricity usage was increasing. He noted that Ireland had a plentiful natural supply of electricity and preference should be given to it. He suggested (without clearly stating) that electricity was the subject of his address, but he wandered far from this topic at times.

The Shannon scheme was nearing completion. It had taken five years and considerable expense, but was expected to be capable of supplying all Irish demand, resulting in a large reduction in fuel imports. He discussed the pricing dilemma (price tied to usage – greater usage lower tariffs), then proceeded to list some useful applications of electricity, such as the production of cement.

Buckley then digressed to examine the "Fundamental requirements of life". He began with air and the problems caused by coal fires, then addressed water and mentioned the recent typhoid epidemic that occurred in a provincial town resulting in several deaths, and stressed the need for proper sanitary organisation. Discussing food, he stressed the need for good public health engineering. He made particular reference to dust-borne disease and did not approve of the traditional (then current) methods of removing dust from the roads by mechanical sweeping. He included an interesting table, which contained the number of fatalities arising from different accidents. Buckley made a reference to the benefits of electric vehicles, his reference being brief as he counted it to be a well-developed technology.

He discussed the use of ozone in the purification of air before moving on to discuss water treatment. Whenever filtration was not sufficient, he considered that the simplest method was aeration, as provided for by a fountain at the Dublin Waterworks at Roundwood as far back as 1867. However, if electricity is available other methods can be used – here he gave a detailed picture of the use of ozone to sterilise water.

He noted that ozone could be used at a range of scales from the domestic to the industrial, and that it had a range of other uses including food preservation and accelerated curing of timber.

Aeration treatment, Vartry waterworks, Roundwood

Buckley then returned to his principal topic "Electricity" and continued by discussing rural electrification in some detail in France, Sweden, Germany and Holland. He mentioned the potential for a low tariff for electricity during night hours, and made an interesting remark about fire-fighting, namely that at that time it was common practice to shut off some town water supplies at night, but that electric pumps and water towers would help address the problem.

He next listed potential agricultural uses of electricity including the artificial drying of grain and forage crops that had been harvested early, fodder pulping and lighting in the byre and dairy.

Discussing railways with respect to electricity, he acknowledged that without an increase in population and industry and agriculture there were only two or three short sections of track that had sufficient traffic to justify electrification unless there is 'the invention of a storage battery of small weight and relatively large capacity which can be charged in, say, ten or fifteen minutes (not many years later, Dr Drumm of UCD developed such a motive-power unit). Buckley also stated that electric vehicles were now feasible, there being more than 100,000 electric vehicles in the USA. The final topic covered by the president was suggesting the establishment of electro-chemical industries in Ireland. Here, as before, he was searching for uses to which the surplus energy generated by the Shannon scheme could be used.

In conclusion, Buckley raised the interesting question 'are we on the eve of the discovery (of) how to 'can' or 'pot' or 'tin' electric energy?' He continued with a series of question and observations on the future, including a recognition of the amount of energy in matter if only it could be released. He singled out the Shannon Scheme as different from other proposals because of 'the celerity of its conception, initiation and completion - all occupying the short period of five or six years'.

Joseph Mallagh began his presidential address by reminding practising engineers that the profession of engineering had been developed by the engineering Institutions, and issued a call to arms, asking the Institution's members to ensure that they gave the public a service surpassing that of any other country. His address developed into an impassioned defence, or perhaps more accurately a celebration, of engineering. Mallagh saw engineering as a driver of profound change at all levels and in all spheres.

His address in 1930 was far seeing: he discussed emerging technologies such as voice recording, cinematography and television, before envisaging the development of "machine-made memory", 'a science-created store house of historical incident both accurate and realistic'. Such prescience is less surprising given that Mallagh identified imagination and memory as two essential facets of an engineering mind. He saw these qualities as being equally vital as the store of scientific knowledge and practical experience.

Mallagh also reviewed advances in materials, particularly steel and concrete, the internal combustion engine's replacement of steam, and the typewriter's replacement of the pen, British Standards, and methods of transport. As his address turned to transport, shipping and port design, it began to become more focused. This is hardly surprising given that he was at the time Chief Engineer to the Dublin Port and Docks Board. Mallagh commented in detail on the need to ensure that port facilities kept pace with developments in shipping. He discussed deep quay walls, dredging, warehousing, cranes, plant for handling bulk cargoes, and road and rail connections to ports. He saw airports as a future extension of seaports and envisaged the need for elaborate lights, signals and landing and embarkation facilities.

Mallagh made reference to the report of the Ports and Harbour Tribunal 1930. This report gave the condition and future prospects of twenty-four harbours in the Free State. Because of his connection with Dublin Port, Mallagh initially refrained from comment and simply described the report as invaluable. Gradually though, forgetting his quibbles, he delved into some aspects of the report, particularly the inference that consulting engineers had been charging unreasonable fees. He drew this inference from the suggestion that engineering advice for harbours controlled by commissioners should be provided by the State. He objected further to the suggestion that State expertise should be hired out to smaller harbours at fees 'which are inconsistent with the professional status of these state officials and the rights of outside engineers'. He supported the call for a proper hydrographic survey of Ireland's coasts. He also suggested that Dublin should develop a "free port" along the lines of Hamburg.

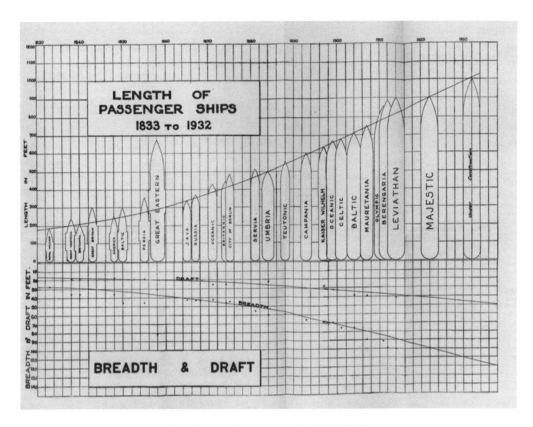

Growth in length of passenger ships 1833-1932

Mallagh included a diagram in his address that showed the increase in the length of passenger ships from 1833 to 1932. His purpose was to show how the development of shipping had initially stalled after, what he termed, "the European war" but that it was starting to recover. However, perhaps the most interesting feature of this diagram is the way in which Brunel's Great Eastern stands out as being revolutionary. The diagram emphasises the fact that this vessel was built literally more than a generation before its time.

Stephen Gerard Gallagher began his relatively brief address in 1931 by reporting that the finances of the Institution were in a healthy state and that membership numbers were also healthy at a total of all categories of 384. Gallagher next discussed the old chestnut of registration and the licensing of engineers. Two years previously the Council had appointed a committee to prepare a draft bill. This Registration Committee, after careful consideration and having received legal advice, had a bill ready for Council's approval prior to it being laid before the Oireachtas. Gallagher was here emphasizing the necessity for registration to protect both the public and the state from unqualified, or semi-qualified people.

Gallagher next addressed the Department of Local Government and Public Health's plans to replace the post of County Surveyor with that of County Engineer. There were concerns that the new county engineers would undertake the design and construction of all new public works. Such a move would have affected the professional practices of members specialising in public works. However, the President informed the meeting that the Minister had clarified the position and that the county engineers would only undertake relatively minor projects. (The comments of both Gallagher and Mallagh reflect the fact that the Institution was concerned at the possibility of the State undertaking design work that had previously been carried out by engineering consultants, the Institution's membership at the time being predominantly comprised of consulting engineers).

Gallagher commented that the rapid development of mechanical transport had led to some difficult problems. He addressed the railway companies' proposals to close a number of unprofitable branch lines and observed that the financial health of some of these lines had never been robust and that their development had been seen as a means of opening up remote areas. He believed that this could be achieved in a more flexible and less costly manner by road transport. However, he called attention to the fact that increased road traffic had led to the need for increased maintenance and investment in better roads.

Gallagher noted that considerable progress had been made in housing and he believed that the bill before the legislators would extend and accelerate this work. He finished his address by acknowledging the need to borrow some of the funds to develop the country's infrastructure. He made the case that there needed to be a balance between the burden on the taxpayers and on posterity because 'if too heavy a burden be placed on the present generation of ratepayers and taxpayers the enterprise and ambition necessary to develop and extend the ordinary business of the country will be cramped and retarded'.

Laurence Kettle began his address in 1933, as had many previous presidents, by addressing the issue of membership. However, he concentrated on emphasising that the Institution included mechanical and electrical engineers within its scope (he himself was an electrical engineer). He observed that the Institution was regarded as being limited to consulting engineers and that some members had in the past not regarded commercial or manufacturing engineers, although qualified, as eligible for membership. Kettle believed it was in the interests of the profession to 'weld the profession in this country into one body', namely the Institution.

On the topic of registration, Kettle reviewed progress. He argued strongly for registration by comparing the numbers of engineers with those in other professions. Unfortunately, the Bill, the development of which had been described by his immediate predecessors, had not been adopted by the government. Kettle acknowledged that the government was engaged in other substantial problems and that this maybe was not the time to press the issue. He believed that Council had done all it could and it was now the duty of members to make further representations.

Kettle's next theme was the need to gather information on Ireland's resources and potentialities. He made particular reference to the valuable work of two previous committees, namely The Irish Peat Committee and The Irish Waterpower Committee. Both committees had been chaired by Sir John Purser Griffith. These committees considered that their work had been merely introductory and the president called for their work to be continued. In particular, he saw the efforts of the Waterpower Committee as having been instrumental in the success of the Shannon power development and transmission network.

Kettle then considered a number of fields that he felt deserved consideration. Amongst these was climatology, where he argued for the State to organise the collection of meterological data, which he considered to be of vital importance for agriculture, forestry, waterpower, drainage and public health. He also called for a thorough survey of the country's mineral resources. On drainage he gave a succinct description of previous drainage initiatives and made a plea for State assistance to complete the arterial drainage of the Irish Free State. He advocated a continuation of the national survey of waterpower and peat resources, and urged a cessation of the wrangling over the development of the latter. He considered that waterpower development should be co-ordinated with water supply needs as 'the storage reservoir which serves waterpower may very well serve water supply also' (Within a few years the ESB and Dublin Corporation developed a joint hydropower and water supply scheme based on a large reservoir near Blessington in county Wicklow).

Speaking of the links between agriculture and engineering, Kettle identified the need for a government department that could give advice, and for mechanical engineers to develop innovative farm machinery. He then considered the use of electricity on farms and identified the benefits of farm electrification. This developed into a more general discussion on rural electrification, which looked at developments in other countries. He thought that in Ireland's case State subsidies would be required.

He then introduced a new topic and tone to his address as he looked back critically on the influence of engineering in the community. He was far from certain that progress had been an unqualified success. He continued: 'The world was probably a much happier and more satisfactory place to live in before the coming of steam power and machinery, and industrial cities and capitalists and communists'. It seems that Kettle had given an address on this theme twelve years before and he quoted liberally from it. The general tone of this section of his address was dystopian* and it is tempting to surmise that it may be the result of the Great War followed by the Great Depression, but his arguments and observations were well made. Kettle's solution consisted of a return to small country industry. He was not a radical and recognised that progress towards resolving society's ills must be gradual and evolutionary.

* The first known use of the word *dystopian*, as recorded by the *Oxford English Dictionary*, is a speech given before the British House of Commons by John Stuart Mill in 1868, in which Mill denounced the government's Irish land policy: "It is, perhaps, too complimentary to call them Utopians, they ought rather to be called dys-topians, or caco-topians. What is commonly called Utopian is something too good to be practicable; but what they appear to favour is too bad to be practicable".

Nicholas O'Dwyer, welcoming visitors to his presidential address in 1934, trusted
> 'that our discussions here will continue to be of sufficient interest to attract the many prominent citizens who have honoured us by coming here, and who have from time to time taken part and added much to the value of those discussions'.

O'Dwyer felt that the 'growing popularity of engineering as a profession is probably due to the mechanical trend of the age in which we live rather than to any commercial urge or to a belief that there is easy money in engineering'. He continued - 'Those most closely associated with the profession are well aware that the financial prospects cannot be regarded as good. In fact they are definitely bad!' He suggested that the engineering degree courses be extended to four years, but that the third year should be spent in an engineering office, e.g. ESB, Ports, Railway, and County Engineer offices.

For the remainder of his presidential address, O'Dwyer turned to the recently passed Act of the Oireachtas, the Town and Regional Planning Act, 1934. The full title of the Act was 'An Act to make provision for the orderly and progressive development of cities, towns, and other areas, whether urban or rural, and to preserve and improve the amenities thereof and for other matters connected therewith'. The Act dealt with inter alia: roads, buildings, amenities, public services, and transport and communication. He dealt in turn with the various parts and sections of the Act in some detail, but endevoured to present the various provisions of the Act in layman's terms. He went on to outline the course that a planning authority and its engineering adviser should pursue in the preparation of a planning scheme.

On 8 August 1935, O'Dwyer chaired a special meeting called to commemorate the one hundredth anniversary of the founding of the Institution. The meeting was convened 'to pay tribute to the foresight of our Founders who met in the Dublin Custom House on the 6th August, 1835, "for the purpose of forming an Institution of Civil Engineers". He addressed the meeting and, on behalf of the Council and Members, thanked the representatives of the various learned bodies present for their support, which he felt 'enables us to more adequately discharge our duty in honouring the memory of our Founders and of the many unselfish and devoted workers who during the past century built up the organisation which we have inherited'. Although O'Dwyer could see little hope of state registration of engineers, he took the opportunity, with government representatives present, of urging 'the recognition of membership of the Institution as a necessary qualification for all state and state-controlled employment of an engineering character'.

In his speech as a visiting dignitary, Sean T O'Kelly, Vice-President of the Executive Council of Saorstát Éireann and Minister for Local Government and Public Health, wished it to be known that the Executive Council was taking a deep interest in the various engineering organisations. He continued: 'The Government, in their efforts to reorganise and carry out reconstruction work in the country, since the foundation of the Free State, were to a large extent dependent on the members of the engineering profession.' Sean T went on to remark that

'Irish Engineering Institutions had produced men who had brought credit and honour to the country'. 'It might be said that in Ireland there were not resources, financial and otherwise, for scientific investigation, particularly in regard to engineering, such as might be possessed by other countries, but notwithstanding that, there was sufficient evidence to show that the Institution of Civil Engineers of Ireland had done extremely well during the one hundred years of its existence'. 'Its members had made themselves famous not alone in Ireland but in all parts of the globe by reason of their remarkable knowledge and wonderful engineering achievements'.

The next president, elected to a two-year term of office in 1936, was **Frank Rishworth**, Professor of Civil Engineering at University College Galway. Having taken leave from the college in 1925 to act as Chief Civil Engineer on the Shannon Scheme, he naturally spoke about his involvement in the scheme up until its completion in 1929. Rishworth referred to the growth in demand for electricity and the ability of the scheme to meet that demand. From the 6th Annual Report of the ESB, he quoted some statistics, including the fact that the plant at Ardnacrusha had provided close on 87% of the total units of electricity generated in the year 1935-36. Moving on from the Shannon Scheme, Rishworth summarized the planning of the harnessing of the river Liffey at Poulaphuca by a combined scheme involving a hydro-electric development and the augmentation of Dublin's water supply.

At the time, a private peat-briquetting plant was just coming into operation in County Kildare and the recently established Turf Development Board (the forerunner of Bord na Mona) was considering a similar plant. Rishworth envisaged peat-fired power stations, but it was to be nearly two decades before the first such station (that at Portarlington) was to come on stream. He paid tribute to the work of Sir John Purser Griffith (the Grand Old Man of Irish Engineering), then nearing his 90th birthday.

The continued growth of Dublin gave rise to traffic congestion, but the solutions had generally to wait until after WW2. Rishworth, in referring to suburban train services, advocated the electrification of the two lines to Bray. He suggested that

'an electric service, with its quicker acceleration and retardation, its inherent cleanliness and possibility of more frequent trains would quickly bring back much of the lost traffic with the aid of bus services at the city termini, whilst additional halts between Dundrum and Shankill would lead to the development of healthy high-lying districts as residential areas now inaccessible to the city worker without a motor car'.

It was to be nearly half a century before this was partially achieved with the introduction of the DART services between Howth and Bray in 1984. The Harcourt Street to Bray service was axed in 1958, but the route was part resurrected in 2004 as the line of the LUAS light-rail service to Sandyford, later extended to

Bride's Glen at Loughlinstown. However, a number of the halts designed to serve housing development remain closed due to the adverse economic factors that prevailed from 2008 onwards.

Speaking of employment prospects for young engineers, Rishworth reminded his audience that at the end of the previous century, the outlook in Ireland for young engineers was very limited, but that the situation had changed dramatically on account of the reconstruction and development policies of the Free State. By 1936 there were plenty of openings for recently qualified engineers. The numerous waterworks, sewerage, road and housing schemes, the activities of the Board of Works on drainage, and the construction of airports, coupled with the developments being undertaken by the ESB, all contributed to the provision of opportunities for young graduates. At the time, some of the engineering schools were becoming overcrowded, entry to the first year classes in the four universities in the State totalling 149 (UCD 53, UCC 35, UCG 46, and TCD 15).

Professor Rishworth concluded 'I have not the gift of prophesy or the imagination of H.G.Wells, but I know the young engineers of today, and I know that they and their successors will not fail, when they have to face the unknown problems, which the future has in store'. At a time when the finances of the Institution were in a healthy state, and during Rishworth's presidency, funding was earmarked for the assistance of distressed engineers and their families, but it was not until 1969 that a Benevolent Fund was formally established.

Joseph Albert Ryan started his presidential address in 1938 by saying that he was sure he would receive the greater measure of support and co-operation from his colleagues, the county surveyors, whom he described as 'the sentinels of the Institution, as well as recruiting officers'. He then read a letter from a member of the Institution resident in England which included the following: 'We must continue to have papers read at our meetings describing all the major engineering works carried out in this country (Ireland) so that the Transactions of the Institution may be a full and accurate historical record of the engineering achievents of this country'.

Ryan continued:
'the public seem largely to regard the work of the Civil Engineer as a sort of high-class tradesmanship, without due regard to the application of science and physics, etc. to the design of his work. I submit that the man in the street who enters a town, say where a sewerage or water scheme is being constructed, regards the trenches in the road as a nuisance and the laying of the pipes as a sort of simple operation which anyone could do, and generally fails to appreciate that behind the scenes there is a directing mind which has spent many laborious hours on the application of the science of hydraulics and mechanics, etc., who has visualised and computed the whole works from beginning to end, has foreseen and overcome difficulties, both physical and financial, and to whom the scheme is a child of his imagination, which he will watch over and guide to its ultimate destination with all the love and care of a fond and loving parent; but to the man in the street it is just another hole in the road'.

'How, then, are we to educate the public to a full appreciation of the value of our professional services? May I suggest that the answer lies with the members themselves - we must tell the public what we do and how we do it – we must let the public know that behind our work there is the application of physicist and scientist, and that we are here to lighten life's burden and to bring all the benefits of modern civilisation to each and every member of the community'.

Ryan felt it appropriate to offer to the members present the following description of Professional Engineer taken from the *Encyclopedia Brittanica*:
'*Professional Engineer:* So diversified are the services required of professional engineers throughout the wide range of industries, public utilities and governmental work, and in the discovery, development and conservation of resources that men of extremely various personality and physique may achieve success. Qualifications include intellectual and moral honesty, judgment, perseverance, resourcefulness, ingenuity, orderliness, application, accuracy and endurance. An engineer should have

ability to observe, deduce, apply, to correlate cause and effect, to co-operate, to organize, to analyze situations and conditions, to state problems, to direct the efforts of others. He should know how to inform, convince and win confidence by skilful and right use of facts. He should be alert, ready to learn, open-minded, but not credulous. He must be able to assemble facts, to investigate thoroughly, to discriminate clearly between assumption and proven knowledge. He should be a man of faith, one who perceives both difficulties and ways to surmount them. He should not only know mathematics and mechanics, but should be trained in methods of thought based on these fundamental branches of learning. Organized habits of memory and large capacity for information are necessary. He should have extensive knowledge of the sciences and other branches of learning and know intensively those things that concern his specialities. He must be a student throughout his career and keep abreast of human progress'.

Ryan continued:

'Having been endowed more or less completely with qualifications and capacities requisite for a professional engineer and having developed them with the aid of educational and other institutions and contacts provided by civilized communities, the engineer is under obligation to consider the sociological, economic and spiritual effects of engineering operations and to aid his fellow-men to adjust wisely their modes of living, their industrial, commercial and governmental procedures and their educational processes, so as to enjoy the greatest possible benefit from the progress achieved through our accumulating knowledge of the universe and ourselves as applied by engineering. The engineer's principal work is to discover and conserve natural resources of materials and forces, including the human, and to create means for utilizing these resources with a minimal cost and waste and with maximal useful results'.

'The wig and gown of the lawyer, the dentist's white coat, the morning-suit of the city man, are that part of the "elevation" which the public appraise highly, and are as valuable in their own degree as a smooth tongue to a salesman or a bedside manner to a medico - and so we Engineers must realise that a suit of plus-fours does not convey the same sense of dignity or professionalism as a morning suit and tall hat. No, gentlemen, I am not advocating the wearing of toppers - but what I do suggest is that all engineers in their dealings with the public must do everything in their power to uphold the status and dignity of the profession. We must educate the public up to our viewpoint, rather than reduce our viewpoint to man-in-the-street level, so as to be on equal terms of understanding'.

The next president, **Henry Nicholas Walsh**, took office in 1940, and will forever be associated with the study of the properties of cement and concrete. His small book, published in 1939, and entitled '*How to make good concrete*', became a best seller amongst the civil engineering profession. As he had spent most of his professional life in engineering schools, it was natural that he devoted most of his address to engineering education and training, dealing with important principles and some of the wider aspects of the subject. Having described in detail the wide range of engineers' interests, he concluded that the first principle of training engineers is to produce men that are versatile, adaptable, disciplined, industrious and healthy. With regard to secondary education, he opined that 'trying to do an engineering course without good mathematical equipment is like trying to climb a mountain in a pair of slippers; it may be done but there will be much slipping and failing and wasted effort'. At the time, University College Cork (UCC) had introduced a special entrance examination in mathematics for engineering students, a pass in Honours Mathematics in the Leaving Certificate being accepted as an exemption.

Commenting on the dropping of French by many secondary schools, he noted that it was a deplorable fact that many students of professional and scientific subjects receive no knowledge of any Continental language. He reminded his audience that 'one of the arguments of Sinn Fein days was that freedom would put us in touch with the Continent. But when the gates of freedom were opened to us the doors of Continental literature were closed to many of our boys'. Walsh considered that it was not possible to keep up-to-date in most branches of engineering unless one read Continental technical literature.

In general, he felt that the completion of a good secondary education should equip a student with mental capacity, character, and sense enough to study for himself and that

> 'the formation of an honest, manly character, of independence of spirit and readinesss to study of his own free will, and to be careful, accurate and thorough in his work, and clear in expression are of far greater importance to a boy than any spectacular success at examinations'.

Professor Walsh was very conscious of the range of abilities shown by his engineering students at UCC, and this is typified by his comment that he had

> 'far more admiration for the man who experiences great difficulty in just passing his examinations than for the very brilliant man who sails through with first class honours. Frequently the former does very well in the outside world because of the stout character that he has formed during his academic struggles'.

He considered that an engineering course was not intended to produce trained engineers or men with wide knowledge of engineering, but that it was intended as a means of training men's minds, through the medium of fundamental engineering theory in such a way that they can assimulate engineering knowledge quickly and that they can develop rapidly under direction when they go into practice.

On the demand for engineers, Walsh traced the increase in demand during the previous twenty years, commencing with the rebuilding of roads, railways and bridges following the destruction occasioned by civil war. The Shannon Scheme and arterial drainage works followed and more road construction. The construction rush began with housing schemes, and the necessary expension of roads, sewers, water mains, more bridges, silos, warehouses, city and county hospitals, and some heavier engineering works, such as the Barrow Drainage, Limerick Dock Extension, Shannon Airport, and the Liffey Scheme. In the period 1925 to 1939, the demand for engineers was very high and seasonally exceeded supply. However, with the advent of WW2 ('The Emergency'), jobs became scarce and unemployed engineers were forced to emigrate. A number of members of the Institution and their families were assisted by the by now well-established Benevolent Fund.

Walsh reported on a joint meeting with the Engineers Association to discuss possible state registration of engineers, expressing the hope that the two bodies might work in harmony and close cooperation. He concluded by answering the age-old question on the lips of potential members of the Institution: what will I get out of it? His stock answer was

> 'What you get out of it will depend on what you put in. If you put in no more than your subscription you will not get much out. If you study the Bulletins and Transactions you will get a good deal. If you come to Dublin now and again to attend and speak at Meetings you will get far more. If you become an active member you will reap a large benefit'.

Thomas Joseph Monaghan commenced his presidential address in 1942 by quoting Robert Mallet's thoughts in 1866 on 'enlarging the engineer's sphere of influence'. Mallet's predictions were deemed remarkable and prophetic and suggested to Monaghan that he himself should present some ideas as to how engineering knowledge may be more fully utilized in future in seeking to attain that splendid ideal of, as Mallet put it, 'comfort, improvement and civilization of our race'. Mallet had said 'who shall predict what new scientific discovery, or the happy application of an abstract truth, new or old, may in a moment unexpectedly call forth'. For example, speaking in 1866, he took an extreme case: 'suppose it were possible that aerial navigation were once (as many fondly imagine) shown to be practicable with advantage, would not the fact instantly call forth a class with the speciality of Aeronautic Engineers'.

In the course of defining the role of the engineer, Monaghan noted that 'while the scientist seeks for knowledge rather as an end in itself, the engineer strives to put his knowledge into tangible shape for the benefit of society'. He felt that engineers should

'educate the public at large to recognize the existence and importance of the profession of engineering; devote more thought to certain aspects, including the financial, of human relationships and reactions; and combat at every opportunity the arguments of those who still assert that one of the conditions of employment for all posts of wide responsibility should be – "No Engineer need apply".

Monaghan continued

'If learning is the hallmark of a profession, then we must point out that the successful practice of engineering today (1942) demands a long and arduous course of mental training, that many of the problems dealt with by the engineer are of a distinctly intellectual type, and that, in so far as the term profession implies intellectual attainments and rigid discipline, we undoubtedly qualify'.

He quoted the saying of Michel de Montaigne, one of the most influential of writers of the French Renaissance: *"I am of the opinion that the most honourable calling is to serve the public and be useful to the many"*. Monaghan, referring back to Walsh's address, emphasized his remark that 'special attention might be devoted to clear and accurate expression in speech and writing', and continued 'it cannot be too strongly impressed upon the young engineer that mastery in the use of language is one of the most powerful aids he can have towards influencing the thoughts, and through them the actions, of other people'.

On the passing on of the experiences of more senior members of the profession, Monaghan wrote: 'There is too much truth for the comfort of some of us in the rather acid saying, *"Experience is the bald man's comb"*, but nevertheless I think that a great deal of good would result if engineers of ripe experience were to read papers on such subjects as *"Some mistakes I have made"*, and *"A few of the wrong turnings I have taken"*, for candid confessions of the difficulties and troubles we have met might remove a few pitfalls from the paths of those who will have to bear the very considerable heats and burdens of the days that are to come'.

Monaghan took time to set down the arguments against the engineer by persons who felt strongly that an engineer should stick to his strictly technical last and that he was not fitted for posts of wide control even over the direct application of his technique, much less for leadership and guidance in non-technical matters. The essence of the case against the engineer was contained in the following opinions of "broadminded administrators", who had studied the problems of human relationships

'while you engineers have been busy with your exact sciences and your experiments, which are so rigid and dispassionate as to unfit you for dealing with the infinite variety of human nature'. 'For example, you announce with pride that "science is measurement". But the things of the mind and spirit, the traditions of the race, the instincts and impulses of humanity, baffle your yard-sticks and balances'. 'Can you measure beauty or justice, "can mathematics give any hints to lovers, is there any scale in physics to assess the aspirations of a saint or the existacies of a poet"?' 'No. Stick to your technical work, at which you are quite good, and we will continue to decide what is to be done with your gadgets and inventions. And as for wider spheres, surely you do not think that ability to adjust even the most intricate physical mechanisms will enable you to handle wisely the vastly more delicate machinery of world trade and international finance. No! the best advice we can give you is not to bother your heads about these abstruse matters; they are beyond your ken. Leave well alone, and control to us, and you will find that everything will be quite all right'. And that, as Monaghan added, 'closes the case against the engineer'.

In presenting the case for, Monaghan felt that

'the world at large was profoundly dissatisfied with the results that have so far been achieved by those who, in the past, have had control over the use and distribution of the engineer's discoveries, and the engineer is confident that he can do better than they have done'.

He continued

'He does not claim that because a man is an engineer he is, by virtue of that fact, a born administrator, but he does assert that, other things being equal, engineering knowledge is a powerful aid in organizing human effort even for non-technical ends'.

Monaghan concluded by noting that

'the outstanding problem that confronts humanity today (1942) is to plot the course that man must follow if hope is not to perish from the earth, and to strive to follow that course. This is a task that demands the utmost efforts by all men of good will and the exercise by them, in patient, persistent cooperation of all the varied talents they possess. The engineer is certain that he has a specially valuable contribution to make towards the achievement of this momentous undertaking, and is determined to devote all his knowledge and skill to it, for he resolutely refuses to abandon the hope that one day all the intelligence, abilities and energies of mankind will be directed mainly towards such noble ends as the "comfort, improvement and civilization of our race" (Mallet, 1866)'.

John Ryan, a past-president, in proposing a vote of thanks to Monaghan, stated that 'it was a pity that the traditions of the Institution would not allow discussion on such an illuminating and educative address which showed such depths of wisdom'.

In his opening remarks, **Thaddeus Cornelius (Con) Courtney** in 1943 recognised that the Institution had been fortunate in having Thomas Monaghan as president during a period when the final drafting of a Bill for the registration of engineers was under consideration and 'his wise guidance did much to smooth away the many difficulties that arose'. Sadly, five years later, the then president, Joseph Candy, would be reporting that, together with the support of the Architects and Cumann na Innealtóirí, a further memorial was being addressed to the Minister for Industry and Commerce reiterating the case for registration of the two kindred professions of engineering and architecture. So the case for registration, which had been first proposed in the 1930s, never became a reality and the *raison d'etre* for it was gradually overtaken by events leading up to the Charter Amendment Act of 1960 and the subsequent unification of the engineering profession in 1968.

Courtney maintained that young men coming into the Institution had achieved a higher standard of training than had existed in the past and 'who had, owing to the changing circumstances of this country during the past twenty-one years, greater opportunities of experience than ever existed before'. He noted that entrance to the engineering profession was mainly through the university engineering schools and that the practice of apprenticeship had almost completely ceased. Present-day students of engineering were expected to possess a knowledge of the fundamental principles of engineering science far more extensive than what was required in the past. Opportunities for practical training had also improved in Ireland. There had been unprecedented activity since the foundation of the State in every phase of engineering work. Electricity demand had risen from 61m units in 1930 to 450m units in 1943. 475 water works and 300 sewerage schemes had been completed and £50m spent on road construction, but the outbreak of WW2 had the effect of postponing many other urgent public works.

At this juncture, the term of office of President was changed from two years to annually.

Liffey Bridge, Dublin: Erected 1816

Taking over from Courtney, **Patrick Joseph Raftery**, giving his address on 6 November 1944, felt that the problem of delivering a presidential address would be made easier if it were to come at the close of his year in office, when part could be devoted to a review of the events of the session.

Noting that it was now one hundred years since the establishment of the Institution of Civil Engineers of Ireland (ICEI) in 1844, Raftery chose the occasion to give an account of the activities of the Institution from its foundation as a Society in 1835, the establishment of the ICEI in 1844, the Royal Charter of 1877, and through to 1944, 'a story of 100 years told in less than 100 minutes' as Raftery put it. Much of this historical account is contained within Chapter One or published elsewhere (See Cox, R.C. (2006) *Engineering Ireland*). Raftery concluded by hoping that the story

'will convince engineers in other lands that they should esteem it an honour and privilege if elected to Corporate Membership of our Institution, and, finally that it will show that Irishmen have every reason to be proud to have here in our capital city, and forever incorporating the name of Ireland in its title, a professional and scientific body of such world-wide influence as the Institution of Civil Engineers of Ireland'.

The next president, **Norman Albert Chance**, began his address in 1945 by acknowledging Raftery's many year of devoted service to the Institution, particularly in the role of Hon.Treasurer. Also, during Raftery's presidency, a record number of 131 new members were elected, and a scheme introduced for inviting members to donate books to the Institution's Library.

Chance's address concerned aspects of civic (sic) engineering in Dublin. He felt nervous in telling his audience of Dublin's past 'lest many of you are not as ignorant of it as I was', and he went on to recall the occasion when he came across an American soldier gazing at Government Buildings and informed him as a courtesy, whilst pointing at 14 Upper Merrion Street, that 'it may interest you to know that the Duke of Wellington was born in that house', to which the soldier replied 'Ah, yes, the old Land Commission – I worked there 25 years ago!'

He then gave an account of the river Liffey and its tributaries, and of the various attempts to bridge the river, beginning with the earliest hurdle bridge located at the end of Bridge Street, the site of the present Whitworth Bridge, during the construction of which a number of earlier bridge foundations were uncovered. Chance noted that Mallagh had previously described in a paper all the Liffey bridges, but singled out the early history of Grattan Bridge and the work of George Semple for further treatment. The first bridge on the site (at Capel Street) was completed in 1678 but, following a number of partial collapses, was replaced between 1753 and 1755 by a five-arch masonry bridge designed by architect/engineer George Semple. Semple was the first to use coffer dams during the construction of the foundations of the bridge down to rock level. He subsequently (1770) wrote a *Treatise upon Building in Water*, regarded as one of the earliest publications on civil engineering, indicating the state of knowledge amongst practising engineers at the time. Semple's bridge, although perfectly sound, was replaced in 1873 by the present structure under the direction of John Purser Griffith, assistant to Bindon Blood Stoney at Dublin Port. The opportunity was taken to accommodate intercepting sewers in each of the abutments.

Chance then presented quite a detailed account of the early state of the approaches from Dublin Bay to the river Liffey, the mapping of the river channels, and the later attempts to define a navigable channel for shipping. He then went on to describe the various land areas reclaimed from the sea and the building of the south and north bull walls.

Turning to Dublin's water supply, he traced the history from the first reliable supplies from the river Dodder at Balrothery in 1244 through to the Vartry scheme of 1865 when the citizens for the first time enjoyed a copious supply of soft and filtered water laid on to every street at high pressure.

To conclude his address, Chance gave a brief historical survey of the sewerage of Dublin and, more particularly, of the various schemes that were put forward for the disposal of the city's sewage. A Royal Commission reported in 1880 and came down in favour of a scheme put forward by the city engineer, Parke Neville, and amended by Joseph Bazelgette. Hassard and Tyrell proposed a scheme that included a one-mile long tunnel to the Nose of Howth, a feature of the later North Dublin Drainage Scheme, completed in 1906. As the proceedings of the Institution contained no account of the main drainage works, Chance gave a brief description of the works, which were constructed between 1896 and 1906. More detailed accounts are available in the Main Drainage Handbook, published in 1906, and in a paper read by Harry Nicholls before the Irish Branch of the Municipal and County Engineers in 1932.

John Purser was elected Hon.Secretary of the Institution in 1939 and President in 1946. Purser's great-uncle was Robert Mallet and he was also related to John Purser Griffith. Purser was a Commissioner of Irish Lights and chose to deal with the work of an engineer in the lighthouse service. He began by tracing the early history of lighthouses and the derivation of the name "Pharos", which became the general name for lighthouses. He explained that Trinity House controlled the lights of England and the Channel Islands, those of Scotland and the Isle of Man by the Northern Lighthouse Board, and those of the whole of Ireland by the Commissioners of Irish Lights. Lighthouses in Ireland were originally controlled by the Revenue Board, but in 1810, control passed to the Corporation for Preserving and Improving the Port of Dublin (commonly known as the Ballast Board).

When the Dublin Port & Docks Board was established in 1867, control of many of the Irish lighthouses passed to the Commissioners of Irish Lights (CIL). Purser described the organization of the CIL, its central stores and workshops and the steam tenders then in service. Of the many lighthouses around the coast of Ireland, none is more iconic than that on the Fastnet Rock off the south west coast of Cork. Purser gave an account of the building of the first lighthouse on the rock, completed by Mallet in 1853, and its later replacement by the present granite tower, completed in 1903, most of the granite coming from Cornwall. During the four years of its construction, there were only 118 days when work was possible, which gives some idea of the difficulties faced by the lighthouse engineer.

Fastnet Rock Lighthouse 1903 (Commissioners of Irish Lights)

There followed a survey of the various illuminants that were used from the earliest bonfires, through the use of coal gas, mineral oil, and paraffin. Early devices for concentrating the light sources into a narrow penetrating beam included reflecting mirrors, but were replaced in due course by lenses, invented by the Frenchman, Fresnel. The lenses (optics) could have any number of sides in order to give a fixed number of flashes per revolution. Purser also dealt at length with communications and navigational aids, including radar, which had been developed during the war that had just concluded and was destined to become standard equipment for shipping and aircraft navigation. At the time of Purser's presidential address, there were 146 lighthouse keepers employed by CIL, now there are none, as lights have become automatic and GPS has made navigation considerable more reliable and safer.

Purser concluded by saying that the work of the lighthouse engineer (at that time) means that he has to be 'a jack of all trades and master of each'. 'He is at constant war with sea and wind and rain and his life is certainly never dull'.

Joseph (Joe) MacDonald began his presidential address in 1947 by opining that
'the Institution has an important duty to perform, which it should not pass over lightly, namely, to encourage education in the broadest sense among its members, foster professional ethics, and promote good fellowship among all engineers. The design, construction and maintenance of engineering works nowadays call not only for the co-operation of civil, mechanical and electrical engineers, but for the specialized knowledge of many professionals, such as geologists, scientists,

chemists, etc., of skilled contractors and manufacturers and, not least, those concerned with the policy of affairs'.

As Chief Civil Engineer of the Electricity Supply Board (ESB), MacDonald chose to speak about the various works with which he had been involved, in particular those concerned with the growth in demand for electrical power between 1925 and 1947. He felt that there should be close co-operation between those responsible for the supply and sale of power and the authorities responsible for the policy of a country.
MacDonald then proceeded to a description of the hydropower stations at Ardnacrusha and at Lough Erne (Cathaleen's Falls and Cliff), the Liffey developments at Pollaphuca, Golden Falls and Leixslip. He also covered the alterations to the Pigeon House steampower station. Demand had risen from 40M Kwh in 1925 to 572M kwh in 1947. Predicting future demand for electricity has always been difficult. MacDonald continued: 'Where the source of supply is partly hydro-electric and partly fuel burning, as in Eire, provision has to be made for dry years with their resultant reduction in capacity of the hydro stations. The total plant must, therefore, cover winter peaks plus standby on these, plus additional fuel-fired plant to cover deficiencies in the hydro plant during dry periods or years.

When considering future needs, MacDonald mentioned small hydro, tidal power and pumped storage. He noted that the tidal range around the coasts of Ireland was considered too small to support a tidal power development of any worthwhile magnitude. With regard to the development of an appropriate site for a pumped storage scheme, he noted that such a scheme must be considered from a variety of aspects, one being the availability of cheap surplus power during nighttime, and noted that storage schemes have an efficiency of the order of 50 to 60 percent.

As previously mentioned, registration of engineers was still a live topic and the incoming president in 1948, **Joseph Phelan (Joe) Candy** believed that the case for registration was a just one and was hopeful that ongoing representations would meet with success.
It had by now become more or less the custom for the president in his address to discuss some aspect of engineering work or engineering organization with which he had been closely associated. Candy, as Chief Engineer of the OPW, chose the subject of arterial drainage.
Referring to the obligation placed on presidents to deliver a presidential address, he felt that this suggested an onerous duty. Michael Mullins must have regarded it as such when his address ran to over 60,000 words and took three evenings to deliver! Candy continued that 'I too have a real appreciation of my responsibilities, but I hasten to assure you that I do not intend to detain you as long as Mr Mullins – in fact, I hope to discharge my obligation within the hour'.

Candy gave a short resumé of arterial drainage work in the past, what the OPW were currently doing, and the problems that lay ahead. He mentioned the previous drainage Acts of 1842 and 1863 and the difficulties encountered, and the fact that many of the districts formed under the Acts had by 1922 fallen into a sad state of disrepair. Candy summarized the earlier attempts at arterial drainage when he mentioned that, 'though the operations carried out under the Acts of 1842 and 1863 are described as works of "arterial" drainage, none of the schemes undertaken was arterial in the true sense of the word; stretches of main rivers and tributaries were taken and cleared, obvious obstructions were removed, the channels widened and deepened, in places re-aligned and possibly embanked, but in no case was a whole river opened up throughout its course. It follows, then, that while considerable benefits were given to lands immediately bordering the portion of the rivers improved, in some cases lands further downstream not previously affected by flooding … were adversely affected as a result of the works upstream. The set-up of these Acts, however, which placed the whole cost directly on the lands benefitted and which required the assent of the majority of the landowners made arterial drainage in its true sense impracticable.'

In consequence, the Arterial Drainage Act of 1925 was passed, this Act reversing the position as it existed under the 1863 Act in that responsibility for designs, estimates and execution was again placed with the OPW. The Barrow Drainage (Act of 1927) was the only scheme to be carried out on anything approaching true arterial lines, and led to the introduction of mechanical excavating plant to the country. With the introduction of mechanical excavating plant, it soon became clear that such methods were no longer the panacea for the cure of unemployment as had been the case in the past. The Arterial Drainage Act of 1945, then recently passed into law, was the first true arterial drainage Act, its most important features being that all arterial drainage work was to be carried out by the OPW at the state's expense, the OPW was to be responsible for subsequent maintenance, and the maintenance costs were to be borne by the county councils as a county-at-large charge. In closing, Candy made mention of possible future catchments destined for schemes of arterial drainage, and the engineering involved. By way of example, he cited the Brosna Scheme, which it was estimated would involve the rebuilding, underpinning, or repair of upwards of 500 bridges in the catchment area.

Michael Anthony Hogan, well known for his occupancy of the chairs of mechanical and civil engineering in University College Dublin, began his address in 1949, like Candy, by referring to Mullins's monumental address in 1860, which opened with a 270-word sentence and ended with an apology for its inadequacy! Hogan's address unsurprisingly dealt with engineering education and research, but also with standards, and co-operation with other professional societies.

On university education, he felt that a professor could not take refuge in an "Ivory Tower", but must keep in touch with the everyday practice of engineering if his school was to keep up to date. Like many of the later educationists, Hogan recognised that the university should not attempt to train an engineer, but rather guide the student into developing an analytical method of approach to problems – a habit of mind rather than actual skill in any particular branch. He noted that, as much of an engineer's time is spent in organization and administration – in dealing with people – some introduction to the problems of management was desirable. He foresaw the university engineering schools moving to a four-year degree curriculum, something that was to become a reality within a decade, but did not agree that such a development was necessary – he felt that a three-year course was sufficient, but supported the idea of one-year post-graduate courses in which advanced topics would be studied.

On research, Professor Hogan said that
> 'it is very easy to adopt a defeatist attitude with regard to research in Ireland, and that, in my view, is quite wrong. It is argued by some that we are a small and poor country that cannot afford to spend money on research, and that we should have the work done by research stations abroad or make use of their researches. One of the troubles with laboratory work done outside the country is the lack of close collaboration that should exist between the laboratory worker and the engineer responsible for the work in hand. In the case of soil samples, for instance, it is very difficult for the laboratory worker to appreciate all the factors involved if he cannot visit the site. There is the further point that the possession of skilled research workers would be a national asset and results could be expected more speedily from a home laboratory. Further, it generally requires a research worker to appreciate fully the significance of results obtained elsewhere, and to form an opinion of their applicability to other conditions. It is not intended to disparage the good work which has been done in connection with many civil engineering projects by outside laboratories, but to say that wherever possible research workers should be trained and laboratories set up to do the work in Ireland.'

Bearing in mind the amount of form filling that is required of a researcher today, it is interesting to note Hogan's remarks regarding administration of research. He said:
> 'From a long experience in engineering research, I consider that two things are essential if research work is to be carried on successfully; first class workers and a sympathetic administration. From the nature of things, a good research worker does not easily fit into a pattern, nor is he good at dealing with forms and it is most important that the organization should provide someone in the nature of a buffer who can deal with accountants' queries and the like, and satisfy them that money has been or will be properly spent, without disturbing the researchers'.

Hogan reminded his audience that the Institution had long favoured the introduction of Standards for Engineering Materials. However, it took some sixty years from the time that John Purser Griffith first pointed out the advantages of adopting a standard specification for Portland Cement before the Institution (in 1938) prepared and issued such a standard in 1938, coincidently in the same year that Sir John passed away at the ripe old age of 98.

The next person to take on the role of president of the Institution probably needed little intoduction to the members present. **Thomas McLaughlin**, regarded as the champion of the Shannon Scheme, had a long association with the Electricity Supply Board (ESB). McLaughlin began his address in 1950 by noting that the pattern of membership of the Institution was changing and he considered that 'there is no doubt that, in the ordinary course of events, its activities will be so broadened that for purposes of papers and discussions it will divide up into many specialist groups, all owing a common allegiance to one professional Institution'. He further noted that the president the previous year was a mechanical engineer, he himself was an electrical engineer, and the following year the president would be a civil engineer.

Speaking about a perceived drift in the concept of professions, McLaughlin referred to a presidential address given by Professor E.B.Moullin, professor of electrical engineering at the University of Cambridge. Moullin felt that Western Europe was veering away from the ideals of the professions and tending towards centralized administration.

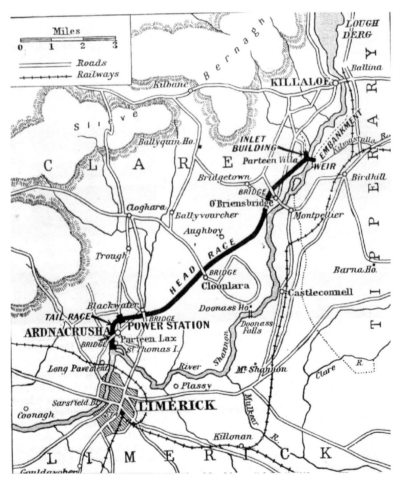

Location of Shannon Hydroelectric Power Scheme

McLaughlin noted that this drift had been apparent in Ireland over the previous twenty years. He was concerned that rather than the concept of the free professional man, the call might be for technical specialists to act as advisers to administrative boards (the present day role of senior engineers in local authorities?).

He remarked that the professions were essentially autonomous bodies, but that the growth of control by central government was tending to make the practice of professions more and more subject to the administrative civil servant who is employed to take decisions subject to the responsible minister and to see that these decisions are carried out.

McLaughlin was led to comment: 'It seems to me this gallant little nation of ours is well to the fore in veering away from the concepts of a free society. However, it is a very melancholy reflection and it leaves me … feeling very uneasy for the future'. Dr McLaughlin concluded his address with these words:

> In these critical days it behoves engineers to give thought to the ideals which created and inspired their profession and gave life to this Institution; to ask themselves is the whole concept not a great and noble one and worthy of their most active support. To the young engineers in particular, I would say, do not fail to realize that without association in an Institution of this nature, there can be no profession. If you agree that professionalism is something well worthwhile preserving, then give every support you can to your Institution, in its activities and in the promotion of membership'.

Following McLaughlin it was the turn of **William Ian Sidney Bloomer** to serve as President. He had served some twenty years in the Department of Local Government, the central authority in Ireland responsible for roads and bridges, waterworks, sewerage, state subsidized housing, town planning, and many other matters related generally to public health and amenity. He chose to review the work of the department since the inception of the State.

He began his address in 1951 with a reference to the public road system in the Republic, which by 1951 comprised 9,839 miles of main road, 39,071 miles of county road and 917 miles of urban road, a total of close on 50,000 miles, an abnormally large mileage compared to other European countries. The roads were in a bad condition in 1922, due partly to neglect during WW1 and partly to Ireland's fight for freedom during which many roads and bridges were disrupted. The usual method of surfacing, even for main roads, was, at the time, waterbound macadam, most of it unrolled and consolidated by traffic. The first phase of road making continued through the years until the major arterial and main roads could be said to have sound waterproof and dust-free surfaces. Many damaged bridges had been restored or rebuilt and it became possible to travel from end to end of Ireland in reasonable comfort and safety if not at great speed. By 1934 the engineers and authorities concerned could look back on a decade or so of fine achievement. Some £100 million was invested and thousands of miles of road improved to the point when the country could be said to possess a sound main road system.

Turning to a review of water supplies, Bloomer noted that all the cities and larger towns now possessed piped water supplies, but that the majority of the smaller towns still depended upon 'the good old pump'. He felt it only proper to note that since the State was established many hundreds of wells, both shallow and deep, had been constructed to supply the rural population. The increasing network of rural electrification was destined to provide a source of power for the operation of pumps and other plant associated with rural water supply schemes.

The policy in Éire at the time in regard to sewerage schemes was that sewerage follows water supply. A survey of existing sewerage schemes in 1947 revealed that Dublin was the only city that had constructed sewage treatment works and that one-third of the total sewage flow of Irish towns was discharged crude to sea, river or lake.

The passing of the Town and Regional Planning Act in 1934 had been a great step forward, but planning authorities were not under any legal compulsion to adopt the measures laid down in the Act. Nevertheless, the effect was to secure more orderly development by speculative builders and other developers and to make provision for open space and adequate space for schools, churches, shops and social services.

Bloomer concluded with these words:
> 'The works of road making and of environmental sanitation never cease. They cannot cease while a nation lives, and their advancement is with good communications, a nation should prosper and progress in peace and contentment'.

Patrick George Murphy, who became president of the Institution in 1952, departed somewhat from previous address formats by, in effect, presenting a paper entitled *'A survey of electricity supply in Ireland'*. Murphy's paper is a valuable account of the founding and development of the Electricity Supply Board (ESB) and its role in the progress of the State. The Electricity Supply Act of 1927 made provision for 'the organization and regulation of the generation, transmission, distribution and supply of electricity' and the ESB constituted to operate the Shannon Scheme and to reorganize the electricity supply of the country on a national basis.

The comprehensive account includes details of the growth in electricity demand and supply, snapshots of each of the generating stations, including the latest turf-fired stations at Portarlington and Allenwood. Murphy used detailed charts to explain the variations in demand and generated units throughout the years from 1931 to 1952 and typical daily load curves for summer and winter. Next, he discussed the high-tension transmission system, the utilization of electricity and rural electrification. Comparisons were made with electricity production in other European countries and remarks about coal production, both in Ireland and imports. Murphy ended with a look at future generating options and a review of the generating plant then under construction.

The Institution continued to grow in membership. **Henry Nicholas (Harry) Nicholls** began his presidential address in 1953 by alluding to the increase in corporate members from 376 to 859 in the thirty years that he had been a member. However, he felt that, 'while the growth in membership is pleasing, we must realize that the Institution needs all the force and strength it can command if it is to combat successfully the present tendency to lower the status of the engineering profession'. Referring to Dr McLaughlin's comments on the gradual exclusion of engineers from administrative positions and their relegation to the position of technical advisors, Nicholls pointed out that the tendency was apparent in many countries, 'but I fear that here it has been adopted as a set policy'. He continued: 'It has also been accompanied by a policy of lowering the salaries paid to engineers. This must inevitably have serious consequences, not only for the members of the profession, but for the country as a whole. Young engineers will not see much prospect of advancement here and will tend, more and more, to emigrate, and we may find in later years, as a result of this policy, the standard of the engineering services lowered'.

Harry Nicholls chose to deal with the branch of engineering that had been a feature of most of his professional life, namely the drainage of Dublin and district. As the Dublin Main Drainage Works, completed in 1906, had never been written up as a paper to the Institution, he took the opportunity to give an account of the works. He noted that subsequent investigations had confirmed that the drying of sewage sludge for manurial purposes was not an economic proposition in a country where the supply of ordinary farmyard manure was plentiful. Nicholls continued with descriptions of the various extensions and additions to the main drainage and made reference to the North Dublin Drainage Scheme then being designed and shortly to cvommence construction.

When the original Main Drainage Scheme was designed the data available to the engineers was very scanty and no records of rainfall intensity existed. Planning had not arrived and the nature of future development was rather vague. Under these circumstances, Nicholls was moved to pay tribute to the men who designed the scheme.

Having dealt briefly with the past and the present, Nicholls made some remarks about the future, in particular in relation to the ever-expanding city of Dublin. He continued:

> 'The increases which have taken place in the past fifty years in the area to be drained have been a matter of concern from the engineering point of view. Any further increase would create serious problems and involve heavy expenditure. Apart from the purely engineering problems, I believe this continued "sprawl of the city" is not in the best interests of either Dublin or the country as a whole and I hope we shall soon see a definite limit fixed and the Green Belt become a real green belt'.

For its next president, the Institution in 1955 turned south and elected **Stephen William Farrington**, City Engineer and Borough Surveyor of Cork and a native of that city. He chose as a motto for his presidential address: "All kinds of men degenerate in secrecy, even as plants blanch in the dark".

Having been in charge of engineering work just prior to WW1, Farrington observed that the war had resulted in 'a change in our social economy from the intense free competition of the nineteenth century to the managed economy of the twentieth'. He continued: 'the change might be summarized by saying that in the nineteenth century, a man invested his own savings; but in the twentieth, the State takes them from him and spends them for him'.

Daly's Bridge in Cork (1926)

Farrington pointed out

> 'that while the change has been effective in doing away with the terrible cruelties that ruthless competition inflicted on the weak, it has in it the seeds of decay, of which the depreciation of the man of action is only a symptom; it has turned man's eyes away from the fact that we live by exacting our requirements from Nature, not from the State. It tends to exalt the the qualities of astuteness and plausibility above those of judgement and action; it tends to standardize the individual and to abolish personality; it is the reason that the man who can do things is less appreciated than he who can talk glibly about them'.

Although he was speaking in the context of the perceived depreciation in the status of the professional engineer, Farrington's portrait of a civil servant is of interest, but might seem a little harsh to our eyes. He considered that civil servants were

> 'men who had passed a written examination with distinction, in early manhood, and thereafter led a life sheltered, regimented, secure, governed by rules, safe from external criticism, sacrosanct even from the indignity of being called to give evidence in a court of law. They acquire a characteristic mentality, address and appearance and above all a characteristic style of writing. They are of unimpeachable integrity, of cast-iron conservatism; they are self-controlled, suave

and courteous; they are admirably equipped to dampen ardour, to quench zeal and to stifle enthusiasm; but they cannot lead'.

Farrington continued:

The civil service can collect taxes so long as they have enormous arbitrary power in their hands, which power they use with admirable constraint; but they cannot secure co-operation, they cannot persuade, they cannot induce. They have not got the necessary background and training; they just do not know how'. 'He who can get things done is he who has been trained to do things himself'. 'It is obvious that as the expenditure of public money tends to become secret, it tends to be transferred out of the hands of practical men into the hands of clerks'. 'The two things go together. If you have not to justify your decision in public, it is not necessary to know the arguments on which it is based'.

Nowadays, we have developed the concept of 'transparency' in matters of state affecting our lives and an expectation of 'freedom of information'. In the mid-1950s, Farrington felt that the status that engineers had lost during the preceeding thirty years was not status as technicians, but 'status as public men'.

Three former presidents had held the position of Chief Engineer at Dublin Port: Bindon Blood Stoney, John Purser Griffith, and Joseph Mallagh, and, if he had not died at an early age, **Francis Willoughby Bond**. Bond was vice-president from 1950 to 1953 and would have been president in 1953-54. The incoming president noted that 'through his untimely passing the Institution was denied the services of one who, without doubt, would have brought to the Chair all those qualities which go to make the ideal president'.

Cornelius John Buckley, then Chief Engineer at Dublin Port, was elected president in 1956. He began his presidential address by calling for an increase in membership and went on to stress the desirability of engineers interesting themselves in other subjects, such as management, finance, economics and administration. Buckley attempted to define the engineering approach to problems as follows:

Clear definition and understanding of the problem in question;

Definition and study of all the factors upon which the solution to the problem must be based and by which it can be affected;

Production of a trial solution;

Meticulous critical examination of this to discover its weaknesses and to devise improvements; and

Repetition of this process until, apparently, the correct solution has been obtained'.

Buckley continued:

'If the problem is a technical one, there follow design, organization, planning and execution with their attendant problems, administrative, financial, legal, staff, labour and so on'. 'Other, and even more important, things that engineering training and experience provide are a sound knowledge and understanding of mankind, a tolerant appreciation of other people's opinions and a constant sense of the vital importance of professional integrity and of honesty and justice in dealing with all those whom one has to work'.

Buckley then turned to the main subject of his address, namely engineering work completed in Dublin port since 1935. In the Transactions may be found, as a portion of his presidential address, a review of improvements and engineering works that had been carried out in the port from its early days up until 1935. Amongst the major works completed since 1935, Buckley singled out the Tobacco Warehouse at the Custom House Docks, and the reclamation of lands in order to build an oil refinery, a project that was subsequently abandoned in 1939.

Cargo Handling at Dublin Port in the 1950s

In order to provide for extensions of the port area, a retaining wall of small concrete caissons with mass concrete superstructure was built on the foreshore of the estuary of the river Tolka and filling of Dublin's refuse placed behind it, a process that proceeded continuously from 1931 until 1954, yielding some 67 acres for port development. The extension eastwards of Alexandra Quay was completed in 1940, and the Ocean Pier completed by 1954. The land originally reserved for a proposed oil refinery was later used to provide for greatly increased oil storage facilities. Between 1951 and 1954 a new graving dock was completed with a view to attracting larger vessels to Dublin for repair work.

Buckley concluded by remarking that '… the port has been well laid out, well constructed and well equipped and, through its engineers, has pioneered a number of constructional methods, a fact in which it may well be allowed to take a small amount of legitimate pride'.

The president in 1957, **Edward Joseph Francis Bourke**, known to his colleagues as 'Ned', was Dublin City Engineer. However, Bourke chose not to describe his work with Dublin Corporation, but rather selected 'The Place of the Engineer in Society' as the subject of his address.

Bourke felt that the Institution should broaden its aims and should concern itself more directly with the national wellbeing, and that it should try to secure its more rapid improvement by engineering techniques. He considered that 'the betterment of national wellbeing now depends more than ever on progress in engineering'. He continued: 'The engineer is the one individual in a society who has been trained to study prime movers, the development of power and the application of forces to the materials of nature, towards the production of greater wellbeing by material means'.

Drawing comparisons with Britain as to real income and wealth, he first noted that it was the free development of engineering that brought about the upward movement of wealth in Britain at the end of the eighteenth century. He felt that economic historians writing of the period were inclined to overlook the thinkers, engineers and technologists whose liberated efforts made possible the accumulations of wealth. In the nineteenth century there was a respect for the technique of engineering that did not and does not exist in Ireland. Not until WW1 was any obstacle placed on free engineering enterprise.

Secondly, Bourke referred to what he described as 'the ill effects of modern governments on wellbeing' and quoted a phrase from an encyclical issued by Pope Pius XI that 'social life has lost its organic form'. He considered that 'as the State thus establishes its monopoly, the citizens grow apathetic and egoistic, cease to cooperate in making social life run smoothly, and finally resist as far as they can all efforts of the central government'.

Referring to the perceived difficulties of democracy peculiar to Ireland, Bourke's considered opinion was that ' in the way democracy works in this country, our influence (as engineers) is quite ineffective in the direction of national affairs'. 'A large proportion of the intelligence and talent of this country has been directed into channels too indirectly associated with the business of increasing material wealth. Our educational system has been directed more towards the accumulation of facts than to the development of thought and initiative; and we place overdue weight on literary achievement and the value of the humanities as a part of the wellbeing of the body politic'. Bourke observed that two-thirds of the newly graduated engineers were leaving Ireland to add to the wealth of other countries and that Ireland was being starved of engineering techniques that could be applied on wealth-producing projects if it were not for the formidable forces opposing such expansion. He felt that the failure of government to embrace new scientific discoveries was due to the lack of scientific and technical capacity in the administration.

Bourke concluded by asking his audience to 'remember that it was an engineering scientist who in recent weeks brought thousands all over the world out of bed at early dawn to watch his creation swimming in the skies with all the hopes and fears his satellite may bring'. He was, of course, talking about Sputnik 1, the first artificial earth satellite, which had been launched by the Soviet Union into an elliptical low earth orbit on October 4, 1957.

The next president, **Jeremiah Augustine O'Riordan**, had spent practically the whole of his working life on the development of electrical energy from waterpower and solid fuel. However, he chose not to devote his address to that topic, but rather to a discussion of some aspects of the engineering profession in Ireland with particular reference to the position of the Institution in relation to it. This was because, during his years as Hon.Secretary of the Institution, he had become closely acquainted with the functioning of the Institution.

He began his address in 1958 by noting that the total number of professional engineers practising in the Republic of Ireland in 1956 was about 1,900, of which about 60% were civil engineers, and 40% mechanical or electrical engineers. It was estimated that not more than about 40% of the total were members of the Institution, but some two-thirds of civil engineers were members. The annual output of engineering graduates (about 160) in proportion to the population was similar to the UK, but as this was in excess of the requirements at home, two-thirds (according to Bourke) of our graduates were being forced to emigrate. O'Riordan felt that the Institution must not overlook this fact and that it had a responsibility towards those of its members who were working in other countries as well as to its home membership. He alluded to the considerable volume of opinion amongst the membership of the Institution that it was not playing as full and active a part as is should in the life of the nation or indeed in the leadership of the engineering profession itself.

O'Riordan regretted that many mechanical and electrical engineers in the country were not members of the Institution, but were members of the Irish branches of the Institution of Mechanical Engineers or the Institution of Electrical Engineers. He suggested that mechanical and electrical divisions should be established within the Institution and also a number of regional branches. It had been decided to form a soil mechanics division within the Institution, the first engineering division to be formed for the promotion of the study of a particular branch of engineering. He noted that around 90% of civil engineers in the Republic were employed by state, semi-state organizations or by local authorities, but he felt that the State had failed to give the Institution the recognition it merited when recruiting professional civil engineers in the public service. He also suggested that moves should be made to achieve recognition of Irish engineering qualifications by the international community (reciprocity of engineering qualifications was destined to come much later under the Washington Accord).

Vernon Dunbavin Harty, the chief civil engineer of the ESB, became president in 1959. He had for some time been interested in the history and aetiology (study of causation) of engineering science and considered that 'an understanding of this history of our science may give us a new interest in our profession and enable us to widen our field of thought for the future'.

Harty proceeded to give a brief history of the engineering profession and attempted to show its evolution from a craft to a profession in which the art and the sciences are closely allied. This evolution was also described by De Courcy when contributing to *Engineering Ireland*, published for the Irish Academy of Engineering in 2006. Harty traced the growth of Egypt as a civilization and that of the Greeks and Romans.

The planning and construction of harbours, transportation systems of road and bridges, water supply and aqueducts, and the development of theories in mechanics and hydrostatics. The Romans, in particular, were great road builders, but it has been said of their bridges that they were like dams with openings in them rather than bridges. Following the fall of the Roman Empire in the fifth century AD, we had to wait nearly a thousand years before the standard of construction of the Romans was again attained in Europe. The Romans had added little or nothing to basic theory, but their construction techniques formed the basis for medieval engineering, such as the arch and vaulting.

Dealing with the Renaissance, Harty dealt with the writings of Leonardo da Vinci (1452-1519), the most versatile philosopher of the period and Italian leadership in engineering construction. Galilei is regarded as the founder of the study of the theory of strength of materials, Hooke produced his famous law 'stress is proportional to strain', and then, of course, there was Newton.

In the seventeenth century, France led Western Europe in engineering skill and this led in time to the establishment in 1747 of the École de Ponts et Chaussées. The building of bridges with segmental arches allowed for a great increase in available waterway under the bridges, a vital contribution to the control of river regimes. The work of Chézy renewed interest in hydraulics and Coulomb evolved a method of determining earth pressure on retaining walls. In 1826 Louis Navier, a professor at the École gave a series of brilliant lectures on applied mechanics and structural analysis that became better known in the English-speaking world following the publication in 1862 of a series of comprehensive textbooks on the subject.

The Industrial Revolution marked a turning point in the technological development of western civilization. France was left behind technologically and Britain literally forged ahead with steam power and iron production. Harty covered the major developments in the nineteenth century, such as road building, canals, railways, and the ambitious use of new materials, particularly for railway bridges.

'Loop Line' Bridge over Westland Row, Dublin 1880

He concluded his address with these thoughts:
'We have seen that the antiquity of our profession ranks with that of law and medicine and that our contribution to society has been no less. So we can take a just pride in this heritage and in the achievements of those before us. In spite of this, little has been written of the history of engineering compared with that of other professions and most engineers have a very casual knowledge of this important subject' 'Much has been said about the humanizing of the training of engineers, and it appears to me that a series of lectures on the history of engineering in our universities would be a valuable contribution to this end'. 'The progress of the modern state is intimately bound up with

engineering in all its forms, but the general public know little of the amount of engineering work which has made such progress possible … and there appears to be a need for a public relations organization to represent the industry as a whole'. 'One or more papers on the history of Irish engineering would also make a valuable contribution to the background material for this publicity'.

Jeremiah Gerard Coffey, then County Surveyor & Engineer for Kilkenny, offered a broad review of the engineering services provided by local authorities in the Republic of Ireland. He classified these services under the following headings: roads, sanitary services, housing, town & regional planning, health services & hospitals, and fire services & civil defence, an indication of the considerable amount of engineering involved in local authority activities. Coffey's address in 1960 contains a good deal of statistical data of value, not only for decision-making in the 1960s, but as a marker for the future.

On roads, he noted that there had been a steady rise in the number of vehicles registered and that the figure was predicted to double in the ensuing ten years. Arterial roads were inadequate for the volume of traffic and road deaths were at the time averaging six per week. Dual carriageways were found to be three times safer than single-carriageway roads. In 1944 the Department of Local Government had issued standards for the classification and layout of roads and these standards formed the basis for the design of road improvement schemes for many years following their introduction.

In general, urban areas were well served with piped water supplies and sewage disposal systems, unlike most rural area where a range of methods of providing water supplies to scattered communities, and minimal provision for wastewater disposal, yet alone treatment, was generally lacking. The growing tendency for the rural population to move to the cities in seek of work was placing a severe burden on the inherited housing stock and new building was not keeping pace with demand resulting in a threefold increase in house prices compared to the pre-WW2 situation.

Coffey dealt briefly with other engineering services before concluding by quoting the Bishop of Galway when delivering the opening address to the first annual conference of Cumann nInnealtóirí held some weeks previously. The bishop remarked that 'this country … depends in a particular degree on the skill and integrity of our engineers'. He continued: 'we in this country could not afford sloth or waste or out-of-date methods. The engineer who gave us the best of modern technology and who ensured by efficient planning and supervision that the people got full and true value was a true patriot and benefactor of our country'.

Thomas Aloysius Simington began his presidential address in 1961 with the humble statement: 'Since I am a contractor, I come to the presidential office, with all humility, through the tradesmen's entrance'. As a consequence, he felt he must be careful to avoid any semblance of advertising in his remarks and was therefore precluded from following the example of his predecessors in office. Instead, he spoke about one of his hobbies, the study of the history of bridge building, which, coupled with his other pastime of trout fishing, enabled him to observe at close quarters many a bridge, and so describe himself as a 'pontist'. In defence of inflicting the subject on his colleagues, he remarked that 'some of these same colleagues have in the past shown scant respect to the splendid heritage of beauty and antiquity embodied in Ireland's older masonry arches'. He gave as examples the water main suspended across the elevation of a beautiful stone arch, or its 'graceful piers underpinned with unsightly blobs of concrete or the mason's craftwork poulticed with mortar, …'

Trim Bridge 1393

Tommy Simington devoted a large portion of his address to a discussion of medieval masonry bridges in Ireland, beginning with the arrival of the Anglo-Normans and the bridges at Dublin and Limerick built at the command of King John around the year 1210. The dating of early bridges still remains a problem, although there are a number of useful indicators, such as the smaller spans of semi-circular or pointed arches, wider piers with cutwaters carried up to parapet level to form refuges for pedestrians due to the narrowness of the carriageway. Simington provided examples of a number of late-medieval bridges of substance and continued with a review of George Semple's work on Grattan Bridge in Dublin and the subsequent publication of his valuable treatise on *Building in Water*, the first Irish textbook on bridge building, which was copiously illustrated with fully dimensioned drawings.

Having mentioned the achievements in bridge building of Charles Vallencey, Lemuel Cox, Alexander Nimmo, and later builders in iron, mainly for rail viaducts, Simington reminded his audience that more than forty papers on the subject of bridges had been presented to the Institution since the Transactions were first published in 1844. These papers have extolled the merits of iron, steel, reinforced concrete, and prestressed concrete in turn to successive generations and Simington confidently predicted that the future would see new materials with characteristics then undreamt of to challenge the skill and ingenuity of the engineers of tomorrow. He considered it safe to forecast that stone would never again compete as a material for bridge construction, despite, as he opined, 'the fact that stone bridges are the only ones that grow old really gracefully'.

He concluded with a plea to those who have responsibility for roads and bridges 'to preserve, wherever it is reasonable and possible to do so, the best of the mason's craft from destruction'.
Simington was later to team up with a later president, Peter O'Keeffe, to undertake the extensive research necessary for their book, *Irish Stone Bridges*, published in 1991.

Tom Simington's successor in office was **Thomas Joseph O'Connor**. His address in 1962 dealt initially with some matters of importance to the Institution and subsequently some aspects of consulting engineering. He had recently represented the Institution at a meeting in London of the Conference of Engineering Societies of Europe and the United States of America (EUSEC) attended by representatives of engineering organizations from Europe, the USA and Canada. In a paper by Professor D.G.Christopherson from the UK, there had been a prediction that professional engineers would be asked to work more closely together in the future. In 1960, the ICEI (Charter Amendment) Act had allowed the Institution to extend its objects and increase the membership of Council and a subsequent Policy Committee agreed that

'in principle the idea of a central, unified, learned society for all professional engineers in Ireland, is logical, and provided that such unity can permit of broad liberty and freedom for individual development of the separate branches of the profession, it must be a good and desirable objective'.

Michael Hogan had previously suggested that, 'if the title of the Institution be changed to "Institution of Engineers of Ireland", it would be all-embracing and more acceptable to engineers, whose main interests lay in branches of our profession, other than Civil'. Examples were cited of the Institution of Engineers, Australia and the Engineering Institute of Canada, which catered for all professional engineers in those countries. O'Connor concluded that 'many obstacles may have to be surmounted before agreement is reached on the best means of achieving unification of all professional engineers into one society in Ireland'.

O'Connor made some observations on the origin and use of the title 'Engineer', generally accepted to derive from the Latin word 'ingeniator' signifying 'a skilled man', and then presented a short historical background to engineering during the eighteenth and nineteenth centuries, in particular in France, as mentioned previously by Simington, and referred to Belidor's five volumes entitled *Architecture Hydraulique*.

Turning to engineering in England, he covered the achievements brought about by or resulting from the industrial revolution, such as improvements in infrastructure, such as roads, canals and railways. He felt that it was reasonable to assume that the pioneer engineers of the industrial revolution in Britain were among the first engineers to practice what is now termed "consulting engineering". Their advice was sought to undertake works where their ingenuity could be of service in the field of civil engineering.

The term 'consulting engineer' was frequently used in the latter part of the nineteenth century, but as the term was undefined, neither a client nor a professional engineer knew exactly what were the duties and responsibilities of a consulting engineer. However, by the time that O'Connor gave his address, the term 'consulting engineer' had a very definite and specific meaning, and in most countries, including Ireland, there was a recognized Association of Consulting Engineers, which in turn, was affiliated to the International Association of Consulting Engineers, better known as FIDIC, which had been constituted in 1913 at Ghent in Belgium.

O'Connor then quoted the FIDIC definition of a consulting engineer as follows:
'A person possessing the necessary qualifications to practice in one or more of the various branches of engineering, who devotes himself to advising the public on engineering matters, or to designing and supervising the construction of engineering works, and for such purposes occupies and employs either solely or in conjunction with another consulting engineer, his own staff, or in the case of a partner or a consultant of a firm of consulting engineers, uses the office and staff of the said firm, and is not directly or indirectly concerned or interested in commercial or manufacturing interests, such as would tend to influence his exercise of independent professional judgement in matters upon which he advises'.

By contrast, **John Lane** in 1963 elected to trace the history of road making from early times to more recent times in Britain and Ireland. He mentioned that archaeologists had unearthed part of a bog road in northern Poland dating from around 500 BC. The earliest example in Ireland is the Corlea iron-age trackway near Keeneagh in county Longford, which has been dated to 148 BC. Even earlier was the road construction system found on the island of Crete, dating from earlier than 2,000 BC. The construction method consisted of the excavation to a width of five yards to a depth of about eight inches, the bottom of the excavation being leveled and compacted. Into this, stones were packed and made watertight with a mortar of clay and plaster. Then came a two-inch thick layer of clay and on this were laid stone flags of about two-inch thickness to a camber that facilitated surface water drainage to side channels of grooved stones, altogether a remarkable example of road building exhibiting quite advanced construction techniques.

The history of Roman road building has been well documented, the method of construction being not dissimilar in some respects to the Cretean method mentioned above. The Romans, of course, were very organized and established a special road construction Corps in the army that was trained in the art of building roads and bridges. To assist with the colonization of parts of Europe and later Britain, roads were built to ensure rapid troop deployments and for communication in general between the major centres of administration. Although the Romans did not extend their road-building activities to Ireland, the country, from early in the Christian era, did possess five main roads, one northwards from Dublin through Drogheda to Antrim with branches to Derry and Ballycastle, another westwards from Drogheda to Roscommon, a third westwards to Clarinbridge (Galway), and roads from Dublin to Limerick and Dublin to Waterford.

Lane next discussed the various methods of construction and surfacing of roads in Britain and Ireland, contrasting the methods of Thomas Telford and John Loudon Macadam, but pointing out that road construction generally made comparatively little progress until the introduction of the motorcar at the end of the nineteenth century. In order to make the roads relatively dust-free, various attempts were made to surface dress the roads with tar, but in 1910 it was recorded that in Ireland there were very few roads with dust-free surfaces. The introduction of road rollers and the increasing use of tarmacadam gradually improved road surfaces to the extent that by the advent of WW2 only some ten per cent of Irish roads remained to be treated. He concluded by saying that he was confident that 'engineers and other scientists will continue to provide from the great resources of road-making materials, highways to satisfy the ever widening demands of man and vehicle'.

At the end of John Lane's term as president, a Special General Committee meeting approved the proposed reorganization of the Institution and its activities set out in the document "Institution a Unified Society Plan for Development". The incoming president, Thomas Kelly, referred to it as 'certainly the most momentous decision taken by the Institution since the Charter was granted almost a century ago'.

Thomas Kelly, like his predecessor in office, John Lane, had been associated for most of his career with local authority engineering. For the subject of his presidential address in 1964 he chose to speak about the changing role of the county engineer and its predecessor, the county surveyor, in the context of roads and the differing views on contract versus direct labour methods of construction.

He began by considering the historical development of the County Surveyor's work from the Grand Juries of the nineteenth century through to the creation of the County Councils under the Local Government (Ireland) Act of 1898. The county surveyors had a rather special status and were expected to tackle the then widespread corruption in the system of roadwork, which at the time was carried out entirely by contract. Their main task, which did not require a high level of engineering skill, was the supervision of a large number of road contracts, to attend the local baronial presentment sessions to meet the local justices of the peace and larger cess payers, to discuss with them new proposals and the condition of existing roads, to certify payments to contractors before the Assizes and to report on those whose work was considered unsatisfactory.

Following the Act of 1898, which replaced the grand juries with county councils, agitation began to replace the contract system with direct labour schemes under the supervision of the county surveyor. The responsibility of the county surveyor was thus greatly increased and in counties where all the roads were taken over by direct labour, his work became very heavy. His clerical staff increased, the motorcar increased his mobility, and the telephone increased his availability.

Kelly dealt in some detail with the various problems facing the county surveyors and county engineers. Brendan O'Donoghue's seminal publication *Irish County Surveyors 1834 – 1944*, provides a comprehensive account of the work of the county surveyors and engineers in Ireland. Following the amalgamation of county engineering services, the broadening of the field of engineering activites afforded the newly titled County Engineer and his assistants a better opportunity for the exercise of their professional skills. The

County Engineer now had full scope for his engineering knowledge, but even more for the exercise of administrative ability because with the years the size of the expenditure controlled by him grew steadily.

Kelly dealt with the effects of WW2 on the road network, the deteriorating surfaces needing to be treated as soon as possible following the end of the war to prevent more permanent damage to the foundations, which to reconstruct would have cost five or six times that of surface dressing. The influence of new road standards, the great growth in motor traffic, and the influx of new ideas from abroad as to methods of road construction began to change the pattern of the county engineer's road organization. The phenomenal growth in motor traffic following WW2, and the trend towards heavier axle loads consequent on the transfer of much of the freight from rail to road, called for the construction of arterial roads, a proportion of which was to be dual-carriageway, but progress was very slow.

Kelly noted that the county engineer needed to develop further the techniques introduced since WW2, including cost analysis, work study, productivity studies, mathematical statistical enquiries, labour relations, traffic studies and planning. These techniques of management, Kelly felt, should not be overlooked when filling the posts of County Manager. He concluded by quoting from a paper on 'The Professional as an Administrator' by L.M.Fitzgerald, Assistant Secretary of the Department of Finance, in which he wrote: 'The contribution that technical people can make to the government of Ireland will continue to grow … the next ten years should see a fusion of professional men and administrators on a scale never before contemplated'. Kelly added 'perhaps ideally that fusion should take place in the one person – the engineer – administrator?'

In 1965, **Richard Ernest (Ernie) Cross** was chosen to be the Institution's 75th president. He was Chief Engineer at the Office of Public Works, where he had close on forty years service. Most of the early members of the Institution were engineers working under the auspices of the Board of Public Works (Ireland), better known nowadays as the Office of Public Works (OPW). A number of the early presidents were members of the Board.

Cross elected to trace the history and development of public works in Ireland before and during the existence of the OPW. This was the first time that the presidential address had been delivered in the Science Buildings of University College Dublin (UCD) in Merrion Street, and now serving as Government Buildings. He entitled his address *"Towards a History of Public Works in Ireland"* and began with a definition of public works as

> 'the duties of creating and maintaining those institutions and public facilities which though they may be in the highest degree advantageous to society in general are, however, of such a nature in extent and cost, that the profit therefrom could never repay the expense to any individual or small number of individuals'.

The Board of Works in the nineteenth century was aptly described by the Allport Commission in 1887 as the "Engineering Department of the Government of Ireland". Public works included harbours, land drainage, irrigation and navigation, roads, bridges, and public buildings.
In his lengthy address, Cross provided the first substantial review of the work of the OPW and it would not be appropriate to reproduce it here to any extent, but rather to mark the important landmarks in the history of the OPW since its foundation in 1831.

Cross briefly referred to public works in Ireland prior to the establishment of the Board of Public Works. He mentioned the development of Dublin Port and the asylum harbours at Kingstown (Dun Laoghaire) and elsewhere on the coast of Ireland, made some remarks on Irish topographical surveys, including the Ordnance Survey, before moving on to roads and bridges. He noted that, apart from post and turnpike roads, private developments had been responsible for certain sections of road construction, and he cited the work of David Aher, engineer to the Castlecomer collieries, who, in 1807, laid out roads from Castlecomer to Carlow, thus aiding the development of the collieries and opening up the surrounding countryside. Canals constructed at public expense, and other inland navigations, notably the river

73

Shannon, were brought under the control of Directors of Inland Navigation, appointed in 1800, and subsequently transferred to the OPW in 1831. Mention was also made of the Bog Commission, set up in 1809 to survey and report on the nature, extent, and practicability of draining and cultivating the bogs of Ireland.

The Board of Works in Ireland (OPW) was established under the Act 1&2 Wm.IV,c.33 on October 15th, 1831, the commissioners appointed being Sir John Fox Burgoyne, the first president of the Institution, John Radcliff, and Brook Otley. Initially, the Board was charged with the execution of the following services:
Management of a fund of £550,000 for loans and grants for public works;
Collection of repayment of advances made out of the Consolidated Fund by the former Loans Commissioners under earlier acts;
Management, maintenance and operation of Inland Navigation;
Control of the fisheries of Ireland previously vested in the Directors General of Inland Navigation;
Certain roads and bridges under the Act 6 Geo.VI,c.101;
Charge of Public Buildings in Dublin and the Phoenix Park;
Administration, maintenance and operation of State Harbours of Dunmore East and Kingstown.
These duties were later expanded and at times contracted as the vicissitudes of the times and the changing political situation demanded.

Dun Laoghaire Harbour (Dun Laoghaire Harbour Co.)

One of the major areas with which the OPW is associated is that of arterial drainage. Although there were a number of earlier attempts at organized schemes of river control, channel development, flood prevention and land drainage improvement, little definite construction appears to have been carried out until the nineteenth century. Cross referred to the 1842 Act and the later act of 1863, both of which were framed to deal with a situation in which the ownership of the large estates affected by the drainage schemes was vested in a comparatively few landlords.
As a result of the preparation of drainage schemes under the 1842 act, it was possible to provide immediate employment in certain distressed areas during the Great Famine of 1846/47.
During this period, the Board of Works made advances for relief work, in some case "not for the sake of the works themselves, but for the relief afforded by them…" The accountants and other Treasury officials at headquarters in Dublin were kept extremely busy, but did not always receive full official recognition of their efforts. In one notable case, the drainage engineer, William Mulvany was made the scapegoat for expenditure overruns on the drainage schemes and relieved of his post. He subsequently teamed up with

a geologist and other experts and together moved to Germany where they initiated a number of deep coal mines that laid the foundation for the Rhur economy. He is remembered by a street name in Dusseldorf.

Cross continued:

'The activities of the Board of Works up to the early 1880s had been a period of vigorous growth in which great energy and resourcefulness was displayed and in which the most diverse public works were carried out embracing practically all aspects of engineering and architectural practice then current. It (the Board of Works) was in effect the principal agent and administrator of the government's efforts towards the economic development of the country'.

As a result of the passing of the Land Acts in 1881, the setting up of the Land Commission and the Congested Districts Board, and developments in Local Government towards the end of the century, the responsibilities of the OPW became very much reduced and this situation remained thus until after the foundation of the State in 1922.

Cross felt that the part played by the Board of Works in the development of Ireland's railway system was not so well known. Its function was to investigate and approve all plans prepared by the railway companies, but also had powers to construct lines in congested districts where a company could not be found to do the necessary work. This led to the construction of a number of light railways, including the West Clare, West Donegal, Tralee to Dingle and tramways such as the Dublin – Blessington steam tramway. The Board's general responsibilities in regard to railways ceased when control of all railways in Ireland was transferred to the British Ministry of Transport on 1 January, 1920. Grants were also made towards the provision of steamer services on the Shannon and connecting coach services. It seems that the attractions and tourist potential of the Shannon region were fully realized, and efforts made to exploit them, long before tourism became "a national industry".

He commented that much had been written concerning the planning and building of the Shannon Power Scheme at Ardnacrusha in 1925-29, but it may not be appreciated that a Shannon Water and Electric Power Bill was introduced in 1899 to 'incorporate a company for the development of water power and electrical energy at Clonlara, a village on the Limerick canal'. The project envisaged certain works at Killaloe, a headrace on the Clare side of the Shannon parallel to the canal, a power station at Clonlara and a tailrace to pass back under the canal to the river. The project received parliamentary approval in 1901 but did not proceed, presumably from lack of public backing and/or financial support.

On 1 April, 1922, the OPW was transferred to the service of the Free State government and its duties performed previously in the Six Counties transferred to the newly formed government for that area. As a means of providing employment, and at the same time developing and improving agriculture, it was decided to reopen works of arterial drainage. A number of acts, including the Arterial Drainage Act of 1925 provided new schemes suitable to the new land tenure conditions and the demands for land improvement.

At a conference held in Ottawa in November 1935, the governments of Ireland, Great Britain and Canada undertook to facilitate the operation of a transatlantic air service and to provide the necessary airports. The OPW surveyed, designed and supervised the construction by contract of the runways and airport buildings at Dublin (1937) and Rineanna (Shannon) (1939).

The work of arterial drainage continued apace and some 26 large catchments had been surveyed and nine completed by direct labour by the time that Cross gave his presidential address.

On the advice of a Swedish expert in fishery harbours, five harbours were selected for development as major fishery stations; these were Killybegs, Castletownbere, Dunmore East, Howth and Galway.

Cross concluded his lengthy address with these words:

'Civilisation as we know it today owes its existence and development over the years to the engineers' and 'The story of the progress of civilization is in effect the story of development of engineering and of the organized national efforts of a dedicated profession'.

The next president, **Daniel (Dan) Herlihy** took up the reins in 1966 at a time when the Institution was going through a period of reappraisal of its role in society, and of the form of organization through which its objectives were most likely to be realized. The simultaneous resignations of the Secretaries to the Engineers' Association and to the Institution presented both societies with an opportunity to serve the engineering profession more effectively. Herlihy then announced that the Council of the Institution had decided to co-operate with the Council of Cumann na nInnealtóirí in establishing a form of joint Secretariat that had a strong appeal to both bodies. More importantly, it had also decided to explore jointly with the Cumann the possibility of unification of both societies.

Herlihy, then Chief Engineer of Córas Iompair Éireann (CIE), took as the topic of his address inland transportation in the Republic of Ireland. He observed that the transportation industry is 'an industry that affects everybody and gives rise to the old tag that nothing is of value until it has been transported'. He reckoned that there were about 600 engineers, or one fifth of all professional engineers in the country, involved directly in the industry and within this group there were at least a dozen engineering specialities ranging from electrical engineers to traffic engineers, all making their contribution to a common end product, the majority being engaged on the improvement and maintenance of the public road system.

Inchicore Works, Dublin

The railways have always been employers of engineers, largely because they designed and built most of their locomotives and rolling stock, and because of the elaborate construction work, and precision, required in railway track and signaling. Herlihy felt that it was not surprising that many engineers who commenced their careers in the workshop and permanent way eventually became managers in the wider sense and that it had been a feature of public transport in Ireland that engineers had become involved extensively in management.

Herlihy noted that the most recent development of interest in the field of transportation was the establishment of the Road Traffic Research Division of An Foras Forbartha (The National Institute for Planning and Construction Research). The concept of having one organization responsible for research on the inter-related subjects of town planning, buildings, and roads was welcomed by the Institution. He accepted that the general set-up in transportation in Ireland, the trends, and the statutory controls and regulations followed the broad pattern of those obtaining in Western Europe, with variations to match the local situation. He noted that railways had lost their former monopoly and were no longer profitable concerns. European governments had, nevertheless, decided to retain railways as a social need and provide them with sufficient financial aid to operate efficiently. A turning point in the history of railways in Ireland occurred in the late 1940s. The cumulative effect of the economic depression of the 1930s and the shortage of materials during WW2 caused railway assets to become run down. Steam locomotives were replaced by diesel locomotive power, and track maintenance and renewal was mechanized. Herlihy remarked that CIE's public transport business by road exceeded its business by rail, the road business falling into four main categories: city bus services, long-distance provincial services, luxury coach tours for tourists, and road freight services. It is interesting to note that Herlihy reckoned average passenger train speeds at 36 to 52 mph and on provincial buses 22 mph, speeds which have improved greatly thanks to the rebuilding of much of the permanent way and the introduction of rolling stock capable of much higher sustained speeds. Long-distance bus services have similarly benefitted from the construction of motorways.

Herlihy concluded with the following words:

'Because of our inability to read the future we cannot be quite sure of the shape of future developments, but we do know that there is plenty of scope for introducing refinements in the motor vehicle and the means of its control while moving through cities. It is an obligation of the engineering profession in this country to keep abreast of these developments so that they may be used to the advantage of the community. This is the practical expression of the advancement of engineering science for which the Institution was established 131 years ago, and I think I can say for the profession as a whole that it is as anxious today as it ever has been to give effect to that objective'.

Patrick (Pat) Raftery began by his presidential address in 1967 by recalling some words of his father when he gave his presidential address in 1944.

'In a Society such as ours it is usually the practice to seek out a distinguished member of the profession and offer him the Chair. I had neither the specialized training nor opportunity to practice to merit the description of a distinguished engineer, but in electing me President of the Institution of Civil Engineers of Ireland, you have made me one and have conferred on me a far greater honour than I could possibly achieve by my own efforts'.

Raftery said that his predecessor in office, Dan Herlihy, deserved special commendation. As a result of his efforts the Institution was now 'a more vigorous body than ever it has been; venerable in maturity, very active and full of vitality, while retaining the distinction of being the second oldest engineering institution in the world'.

Presidents in their addresses had spoken on many matters pertaining to engineering. In the era when technical literature and publications were few it was the custom to review recent advances in engineering and technology, but Raftery felt that 'no contemporary president could hope to survey and discourse on the present-day wide field of technological advances, and that these addresses are now of more interest to the social and economic historian than to the engineer'.

The papers and addresses that have been published in the Transactions evidence the progress of our country since the early nineteenth century. In these will be found accounts of the construction of the railways, bridges, ports, canals, river and drainage works, the reclamation of boglands and the utilization of peat resources, the development of roads and transportation, and the provision of electricity to the country; and the advancement of technical and university education among other subjects. As Raftery observed 'all of which demonstrate most vividly how engineers have contributed to the advancement and enrichment of our country'.

For the past year or so, a Joint Committee, composed of members of the Engineers Association and the Institution had been examining the possibility of having a single society to represent engineers and to promote engineering knowledge and mechanical science and technology. Raftery said that 'it was confidently expected that clear proposals for more effective and efficient ways of promoting engineering knowledge and serving engineers will evolve from these deliberations'.

Following the 1960 Charter Amendment Act, a revitalized Institution had been implementing a number of new initiatives, one of which was the creation of a number of divisions catering for specialized sections of the membership. Another initiative was a decision that issues of importance to the wellbeing of the country and on which engineers could give expert and authoritative consideration should be examined and reported on, and a number of national seminars were forthcoming and reports prepared on the outcome of each.

Raftery felt that the distinction between engineers and engineering was ill-defined, although the division is often apparent. He noted that in this country one organization, the Institution, is concerned with the structure of the profession and its promotion. The Engineers Association is also concerned with the promotion of the profession and extends its activities into the personal problems of individual engineers. Salaries, pensions, appointments and legal problems are all its domain. He reminded the members that 'the standing of the Institution, its legal status as set up by Statute, its powers to qualify engineers, its position as a "legal charity" with the consequent advantages are all matters which must be taken into consideration by the Joint Committee when reviewing the position of engineering organizations in Ireland'.

Having made some remarks about premises and services, the library, and the report on the education and training of technicians that had received the approval of Council earlier in the year, Raftery appealed to and strongly advised every suitably qualified engineer to apply for membership of a professional institution.

The last president of the Institution prior to its unification with the Engineers Association was **James Dooge**. Jim Dooge played a leading part in the merger negotiations with Jock Harbison and others and steered the required legislation through the Oireachtas. For his address in 1968, he selected three topics, related to one another, but each important in its own right and all matters of concern to the Institution: the relationship of science and engineering, the relationship between engineering education and engineering training, and the position of the engineer in society.

The close link between engineering advance and scientific knowledge developed nearly two hundred years ago around the same time as the emergence of engineering as a profession. Dooge quoted from Tredgold's 1828 description of civil engineering, which contained the well-known concise description as "The art of directing the great sources of power in nature for the use and convenience of man", but contained also the following closing remarks:

'... the real extent to which it (civil engineering) might be applied is limited only by the progress of science; its scope and utility will be increased with every discovery in philosophy, and its resources with every invention in mechanical or chemical art, since its bounds are unlimited and equally so must be the researches of its professors (practitioners)'.

Dooge continued:

'Engineering is not just some debased form of science, some half-brother scarcely to be acknowledged by the legitimate heir. Engineering has its own function, its own purpose and its own rewards'. 'We must continue to respect science in its own sphere and to use its results in ours, always with due acknowledgement, but without any apology for our concern with the practical'.

On education and training, Dooge had this to say:

'Modern practical knowledge is so specialized and developing so rapidly that those who retire to the cloister of academic life from the world of engineering practice may soon find themselves teaching the technology of yesterday when they would be better off introducing students to the scientific tools required for the technology of tomorrow'. 'In the past, the engineer needed only to equip himself with a knowledge of the physical sciences of his day together with some elementary economics and to acquire, if it was not innate in him, a certain capacity for the handling of men. Today the engineer has not only to apply a scientific knowledge of great scope and complexity, but he has to apply it to man-machine systems, whose very size calls for new sciences to analyse, design and control them'.

On the issue of women in engineering, Dooge felt that it was no wonder that many people still thought of engineering in the context of manual work, when the proportion of women in engineering was substantially lower than in nearly all the other professions. He continued: 'To my mind there is hardly anything that would improve the public status of the profession more rapidly than a transformation of this most regrettable state of affairs'. Dooge longed to see the day when speakers at the Institution meetings would not, without discourtesy, be able to commence their remarks by saying "Mr President and Gentlemen". He wondered whether it was asking too much to hope that he lived to see the day when they will have to preface their remarks with "Madame President". (He did indeed see the day, as Jane Grimson was elected President of the Institution in 1999).

Jim Dooge concluded his address by referring to wonderful opportunities that would be afforded by the passing of the Charter Amendment Act (1969) and he urged the younger members to grasp firmly the opportunity and asked them to recognize that the new Institution (The Institution of Engineers of Ireland) will have within its power the shaping of the engineering profession in this country during the next quarter of a century. He continued that it was for them to make of the Institution what they will or by their apathy to allow it to decline from its present high level of activity.

As President of the Institution, Dooge, in talking to the younger members, asked of them that they
'take this Institution, and its new charter and shape it so vigorously to their own will that very shortly I myself will not recognize the Institution for what it was. I may then take on the stereotype of a past-President – an ageing reactionary deploring the rashness of the younger generation and lamenting the passage of the Institution as he has known it'.

22 Clyde Road, Dublin: Headquarters of the Institution of Engineers of Ireland (Engineers Ireland)

Presidential Addresses in ICEI Transactions

1856	George Willoughby Hemans	: 9 December 1856: Vol. V, 1860, pp 51 – 65
1859	Michael Bernard Mullins	: 8 November 1859; 13 March 1860; 22 May 1860: Vol. VI, 1863, pp 1 – 186
1861	Richard John Griffith	: 19 February 1861: Vol. VI, 1863, pp 193 – 220
1865	Robert Mallet	: 7 February 1866: Vol. VIII, 1868, pp 48 – 102
1867	William Anderson	: 11 December 1867: Vol. VIII, 1868, pp 173 – 209
1869	John Ball Greene	: 9 February 1870: Vol. IX, 1871, pp 109 – 156
1871	Bindon Blood Stoney	: 12 January 1872: Vol. X, 1875, pp 45 – 64
1873	Charles Philip Cotton	: 11 February 1874: Vol. X, 1875, pp 101 – 123
1875	Alexander McDonnell	: 16 February 1876: Vol. XI, 1877, pp 61 – 95
1877	Robert Manning	: 6 February 1878: Vol. XII, 1880, pp 68 – 85
1879	John Bailey	: 14 April 1880: Vol. XIII, 1882, pp 51 – 72
1881	Parke Neville	: 1 March 1882: Vol. XIV, 1884, pp 60 – 81
1883	William Hemingway Mills	: 7 November 1883: Vol. XV, 1885, pp 1 – 29
1885	John Audley Frederick Aspinall	: 4 November 1885: Vol. XVII, 1887, pp 1 – 18
1887	John Purser Griffith	: 7 December 1887: Vol. XIX, 1889, pp 34 – 78
1889	Spencer Harty	: 3 December 1890: Vol. XXI, 1892, pp 1 – 53
1891	Thomas Francis Pigot	: 13 January 1892: Vol. XXII, 1893, pp 3 – 24
1893	John Chaloner Smith	: 4 January 1893: Vol. XXII, 1893, pp 74 – 88
1896	James Dillon	: 2 December 1896: Vol. XXVI, 1897, pp 40 – 79
1900	Edward Glover	: 5 December 1900: Vol. XXVIII, 1902, pp 98 – 123
1902	John Henry Ryan	: 3 December 1902: Vol. XXX, 1904, pp 10 – 31
1904	Robert Cochrane	: 7 December 1904: Vol. XXXII, 1907, pp 26 – 81
1906	William Ross	: 3 January 1906: Vol. XXXIII, 1907, pp 30 – 37
1907	Joseph Henry Moore	: 6 November 1907: Vol. XXXIV, 1908, pp 1 – 26
1909	George Murray Ross	: 3 November 1909: Vol. XXXVI, 1910, pp 1 – 23
1911	Peter Chalmers Cowan	: 1 November 1911: Vol. XXXVIII, 1912, pp 2 – 26
1913	William Garibaldi Collen	: 5 November 1913: Vol. XL, 1914, pp 1 – 25
1915	Mark Ruddle	: 3 November 1915: Vol. XLII, 1916, pp 1 – 21
1917	Walter Elsworthy Lilly	: 5 November 1917: Vol. XLIV, 1919, pp 1 – 16
1918	John Ousley Bonsall Moynan	: 4 November 1918: Vol. XLV, 1921, pp 1 – 21
1919	Patrick Hartnett McCarthy	: 3 November 1919: Vol. XLVI, 1922, pp 1 – 22
1920	Francis Bergin	: 15 November 1920: Vol. XLVII, 1923, pp 1 – 11
1921	Joshua Hargrave	: 7 November 1921: Vol. XLVIII, 1923, pp 85 – 106
1922	Pierce Francis Purcell	: 6 November 1922: Vol. XLIX, 1924, pp 1 – 42
1924	James Thomas Jackson	: 3 November 1924: Vol. LI, 1926, pp 1 - 18

1926	Arthur Hassard	:	1 November 1926: Vol. LIII, 1928, pp 1 - 31
1927	Alfred Dover Delap	:	7 November 1927: Vol. LIV, 1929, pp 1 - 36
1929	Michael James Buckley	:	4 November 1929: Vol. LVI, 1931, pp 1 - 50
1930	Joseph Mallagh	:	3 November 1930: Vol. LVII, 1932, 1 – 21
1931	Stephen Gerald Gallagher	:	2 November 1931: Vol. LVIII, 1932, pp 3 - 8
1932	Laurence Joseph Kettle	:	7 November 1932: Vol. LIX, 1933, pp 1 - 32
1934	Nicholas O'Dwyer	:	5 November 1934: Vol. LXI, 1935, pp 1 – 21
	Special Centenary Address	:	6 August 1935: Vol. LXII, 1936, pp 1 - 52
1936	Frank Sharman Rishworth	:	2 November 1936: Vol. LXIII, 1937, pp 1 - 17
1938	Joseph Albert Ryan	:	7 November 1938: Vol. LXV, 1938-39, pp 1 - 12
1940	Henry Nicholas Walsh	:	4 November 1940: Vol. LXVII, 1940-1941, pp 1 - 20
1942	Thomas Joseph Monaghan	:	2 November 1942: Vol. LXIX, 1943, pp 15 - 37
1943	Thaddeus Cornelius Courtney	:	1 November 1943: Vol. LXX, 1944, pp 1 - 14
1944	Patrick Joseph Raftery	:	6 November 1944: Vol. LXXI, 1945, pp 1 - 20
1945	Norman Albert Chance	:	5 November 1945: Vol. 72, 1946, pp 1 - 21
1946	John Purser	:	4 November 1946: Vol. 73, 1947, pp 1 - 16
1947	Joseph MacDonald	:	3 November 1947: Vol. 74, 1948, pp 1 - 19
1948	Joseph Phelan Candy	:	6 December 1948: Vol. 75, 1949, pp 35 - 53
1949	Michael Anthony Hogan	:	7 November 1949: Vol. 76, 1950, pp 1 - 15
1950	Thomas Aloyius McLaughlin	:	6 November 1950: Vol. 77, 1952, pp 1 - 10
1951	William Ian Sidney Bloomer	:	5 November 1951: Vol. 78, 1953, pp 1 - 14
1952	Patrick George Murphy	:	3 November 1952: Vol. 79, 1953, pp 1 - 16
1953	Henry Nicholas Nicholls	:	2 November 1953: Vol. 80, 1954, pp 1 - 12
1955	Stephen William Farrington	:	7 November 1955: Vol. 82, 1956, pp 1 - 14
1956	Cornelius John Buckley	:	5 November 1956: Vol. 83, 1957, pp 1 - 13
1957	Edward Joseph F. Bourke	:	4 November 1957: Vol. 84. 1958, pp 1 - 12
1958	Jeremiah A. O'Riordan	:	3 November 1958: Vol. 85, 1959, pp 1 - 12
1959	Vernon Dunbavin Harty	:	2 November 1959: Vol. 86, 1960, pp 1 - 13
1960	Jeremiah Gerard Coffey	:	7 November 1960: Vol. 87, 1961, pp 1 - 19
1961	Thomas Aloysius Simington	:	6 November 1961: Vol. 88, 1962, pp 1 - 11
1962	Thomas Joseph O'Connor	:	5 November 1962: Vol. 89, 1963, pp 1 - 14
1963	John Lane	:	4 November 1963: Vol. 90, 1964, pp 1 - 15
1964	Thomas Kelly	:	2 November 1964: Vol. 91, 1965, pp 1 - 20
1965	Richard Ernest Cross	:	4 October 1965: Vol. 92, 1966, pp 1 - 29
1966	Daniel Herlihy	:	3 October 1966: Vol. 93, 1966-67, pp 1 - 5
1967	Patrick Raftery	:	4 December 1967: Vol. 94, 1967-68, pp i - vi
1968	James Clement Ignatius Dooge	:	7 October 1968: Vol. 95, 1968-69, pp 1 - 7

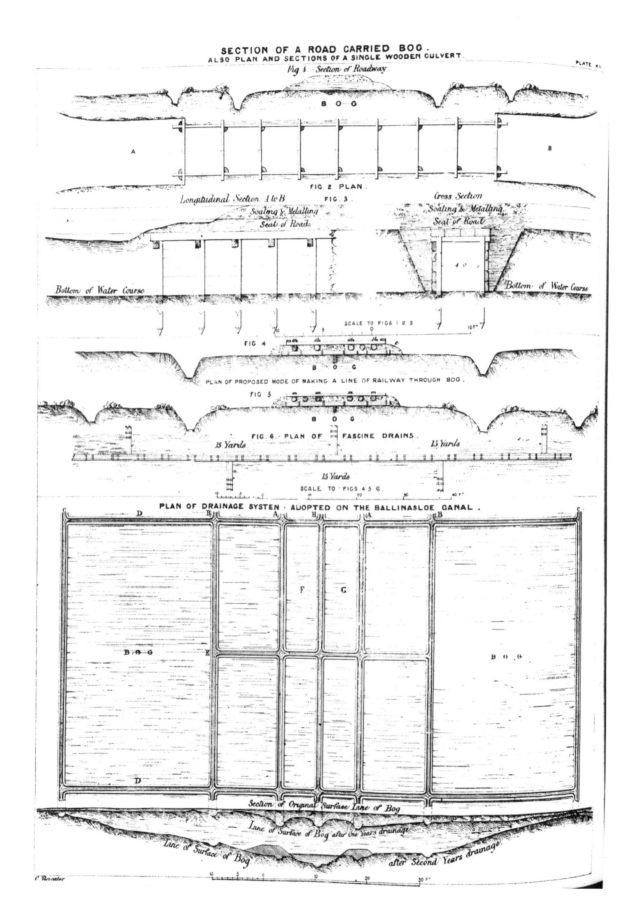

Chapter Three
Biographical Sketches of Presidents

The sketches are stand-alone compilations of the biographical details of each president. There is inevitably some repetition in describing their roles in the Institution and in the content of their presidential addresses.

BURGOYNE, Sir John Fox, (1782-1871), military engineer, was born at Queen Street, Soho, London on 24 July 1782, the eldest of four illegitimate children of Lieutenant-General John Burgoyne, FRS (1723-1792) and the singer Susan Caulfield. Sir John's godfather was Charles James Fox, from whom he took his second name. On the death of his father in 1792, John Fox was placed in the care of his uncle, the 12th Earl of Derby.

Burgoyne was educated by private tutor in Cambridge before entering Eton School in 1793. On 19 October 1796, before he had reached his fifteenth birthday, he entered the Royal Military Academy at Woolwich and on 29 August 1798 obtained his first commission as a second lieutenant in the Royal Engineers. He was to spend more than seventy years in army and public service.

By the end of the Napoleonic Wars, Burgoyne, now with the rank of Lt-Colonel, had been eight times mentioned in dispatches, had received five gold medals, the cross of the Tower and Sword, and the decoration of the Bath, but to his great disappointment missed the Battle of Waterloo.

On 31 January 1821 John Fox Burgoyne married Charlotte, daughter of Hugh Rose of Holme near Inverness, and by her had seven daughters and an only son Captain Hugh Talbot Burgoyne RN, who died at sea in 1870.

Having served from 1821 until 1826 as Commanding Officer of the Royal Engineers at Chatham, and later at Portsmouth, he was seconded from the Royal Engineers and appointed in 1831 to the civilian post of first Chairman of the newly established Board of Works (Ireland), an office that he was to hold for the next thirteen years. Whilst living with his family in Ireland, he resided at Monkstown in south county Dublin. During this time he was engaged on many undertakings of public utility. He acted as Chairman of the Commission for the Improvement of the Navigation of the Shannon, under which extensive engineering works were completed. His experience of the blowing up of bridges in war served him in good stead when it came to removing a number of bridges along the Shannon that were interfering with navigation and drainage. He was concerned to ascertain in great detail how to bring about the most effective and least dangerous detonations (on the intuitive supposition that the cleaner the explosion the more muted the noise) and passed on information on the best quantities of powder charge to use according to the rock type, variations in the diameter and depth of bore to be used, and the methods of tamping, thus applying his military experience to civilian public works.

He was a member of the Commission appointed in 1836 to consider and recommend a general system of railways for Ireland. Commission members included Thomas Drumond, Peter Barlow, Sir Richard Griffith, and Sir Harry Jones, who acted as Secretary. Burgoyne was also at various times Chairman of the Commissioners of Drainage, a member of the Board of the Wide Streets Commissioners for Dublin, and Chief Commissioner for the Royal harbours at Dunmore East and Kingstown (now Dun Laoghaire).

As we have seen in Chapter One, Burgoyne was one of the founders and the first president of the Civil Engineer's Society of Ireland (founded 1835 and in 1844 renamed the Institution of Civil Engineers of Ireland), and delivered an inaugural address in the Custom House in Dublin on 6 August 1835. He organised the examinations for appointments to county surveyorships, established in 1834. He was elected an Honorary Member of the ICE on 12 February 1839. The ICEI conferred a similar honour on Burgoyne in 1845 on the occasion of his resigning the presidency and his planned return to England to take up the post of Inspector-General of Fortifications and assuming command of the Royal Engineers.

However, Burgoyne was not quite finished with public service in Ireland as he was called upon, between February and September 1847, to organise and direct the operations of the Relief Fund for the distress in Ireland caused by the Great Famine 1846-47. Following the famine, during which the population was reduced from eight million to around five million, a 'laissez faire' policy was adopted by the government agencies and engineering works such as arterial drainage came to a complete standstill.

On his return to England in 1848, he was appointed one of the royal commissioners to superintend the completion of the Palace of Westminster and in the following year, was asked to report on the state of the Caledonian Canal and the causes of flooding in Inverness.

Burgoyne received many honours in his lifetime, including being created a Baronet in 1856, a Grand Officer of the Legion of Honour, and being conferred with the degree of DCL (Doctor of Civil Laws) by the University of Oxford. He was elected a member of the Royal Irish Academy in 1834.

After many more years of military service, Burgoyne died at 5 Pembridge Square, London on 7 October 1871 of nervous eczema brought on, it is thought, by the tragic loss of his only son during the previous year. He was buried in St Peter's Church in the Tower of London.

RADCLIFF, John (1800-1869), military engineer, was likely born in England about 1800. Nothing has been found to indicate the nature of his early education before, on 18 January 1814, he entered the Royal Military Academy at Sandhurst as a cadet. He graduated from Sandhurst on 24 July 1820 and was commissioned as a second lieutenant in the Royal Engineers.

It should be noted that from 1821 until 1826, John Fox Burgoyne was commanding officer of the Royal Engineers at Chatham and may have had a part to play in Radcliff obtaining appointments in Ireland and later joining him in the Irish Board of Works (now known as the Office of Public Works or OPW).

On coming to Ireland, probably in 1826, Radcliff was appointed one of four Directors of Inland Navigation and was also appointed a Commissioner for Roads and Bridges. The commission was set up in 1826 to provide for the repair, maintenance and repair of roads and bridges previously built in the west of Ireland under the government civil engineer John Killaly and in the southwest under Richard Griffith.

In 1831, at the age of 34, Radcliff joined Fox Burgoyne as one of two commissioners at the OPW, working from 27 Merrion Square South in Dublin. The work of the OPW was extensive and included public buildings, roads, inland navigation, drainage and even railways. Radcliff would have played a fully supportive role to the chairman, Burgoyne. As civil works increased, the situation was eased by the appointment of a third commissioner, Colonel Harry David Jones, another member of the Royal Engineers, but the activities of the Board were to increase dramatically when faced with the relief of the famine of 1846-7. An important role assigned to the OPW, under an Act of 1833, was the selection of suitable persons for appointment as county surveyors. The selection panel consisted of Burgoyne (Chairman), Radcliff, and Jacob Owen (engineer and architect to the Board) and they carried out a rigorous system of written examinations and interviews occupying on average nine hours a day for a full two weeks.

John Radcliff was a founder member and a trustee of the Civil Engineers Society of Ireland, (renamed the Institution of Civil Engineers of Ireland). He was vice-president in 1845 and 1847, and served as president in 1846 in succession to John Fox Burgoyne. This was the year in which the institution was placed on a firm footing by the introduction of by-laws (largely drafted by Robert Mallet) and the publication of Transactions more befitting a learned society'.

Radcliff retired in 1863 and died at 70 Morehampton Road, Dublin on 21 May 1869 of heart disease.

JONES, Sir Harry David (1791-1866), military engineer, was born at Landguard Fort, Felixstowe, Suffolk on 14 March 1791, the youngest son of John Jones, General Superintendent at Landguard Fort, and Mary Roberts. Harry Jones was one of a triumvirate of British military engineers connected with public works in Ireland who guided the Irish professional institution of civil engineers in its formative years, the others being John Fox Burgoyne and John Radcliff, all of whom at one time served as president of the professional engineering body.

Harry Jones entered the Royal Military Academy at Woolwich on 10 April 1805 and was commissioned a second lieutenant in the Royal Engineers on 17 September 1808.

In May 1835, following an illustrious career in the army, during which he at one time served under Burgoyne, Jones was appointed to his first public office, that of Commissioner for Municipal Boundaries in England, but in November of that year was sent to Ireland to assist the commissioners appointed for the improvement of the River Shannon, a project with which he was to be connected for several years. However, his services were not confined to the work on the Shannon as in 1836 he was appointed Commissioner for fixing the Municipal Boundaries in Ireland and in October of that year was appointed Secretary to the Irish Railway Commission. He also was asked to report on the state of distress in county Donegal.

In 1839, Major Jones returned to the army Corps of Engineers, being appointed Commanding Royal Engineer at Jersey, but was again seconded to civilian duties, being this time made Commissioner for the Improvement of the Navigation of the Shannon. Although he was appointed to the office of Inspector General of Fortifications in 1842, the Lords of the Treasury allowed him instead to continue his work in Ireland and in 1845 he became Chairman of the Board of Works (Ireland) in succession to Burgoyne and worked with him on the famine relief committee in 1846-7. Finally, in 1850, Jones left Ireland to resume his military career in the Corps of Royal Engineers, having, for 15 years, given loyal public service in Ireland.

In 1824 Jones married Charlotte, second daughter of the Revd Thomas Hornsby, rector of Hoddesdon, Hertfordshire.

Jones was elected a Corresponding Member of ICE in May 1837. He was elected MICEI in 1845 and served as ICEI president in 1846.

He took part in a number of military campaigns that took their toll on his health (he was severely wounded in the Crimean War) and obtained many honours, both military and civil, including a Knighthood and the French Legion of Honour. In May 1856, he was appointed Governor of the Royal Military College at Sandhurst, where he died following a stroke on 2 August 1866. He is buried in a vault in the cemetery of the college.

GRIFFITH, Sir Richard John, (1784-1878), civil engineer and geologist, was born on 20 September 1784 at 8 Hume Street, Dublin, son of Richard Griffith, of Millicent, Naas, county Kildare, and his first wife Charity, daughter of John Bramston of Oundle in Northamptonshire.

Griffith attended school at Portarlington and Rathangan before completing his early education in Dublin. In 1800 he became an ensign in the Royal Irish Regiment of Artillery stationed at Chapelizod near Dublin, but, following the Act of Union in 1801, he retired from the army to pursue a career as a civil and mining engineer.

Around 1802 he went to London, where he studied chemistry, geology and mineralogy for two years, subsequently going to Cornwall to study practical geology. He visited Matthew Boulton at Birmingham and also the various mining districts in England and Scotland. Between 1806 and 1808 Griffith studied geology at Edinburgh University and in 1807, at the age of 22, was elected a Fellow of the Royal Society of Edinburgh. It was in Scotland that he met his future wife, Maria Jane Waldie, eldest daughter of George Waldie of Hendersyde Park, Kelso, whom he married in Kelso on 21 September 1812.

Griffith returned to Ireland in the Spring of 1808 and for the next fourteen years devoted himself to an investigation of Ireland's geology, including the search for the mineral wealth, which it was hoped would trigger an Irish industrial revolution. In 1812 the Royal Dublin Society (RDS) appointed Griffith as their Mining Engineer and the same year he was appointed Inspector-General of His Majesty's Mines in Ireland. He completed surveys of coalfields in Connaught, Ulster and Munster, and lectured on geology and mining at the RDS during the period 1814 to 1829, when he resigned his position with the society. During all this time, Griffith collected information for his *Geological Map of Ireland,* the first edition of which appeared in 1815. For this and other like achievements, he was awarded in 1854 the Wollaston Medal of the Geological Society of London.

On 28 September 1809 Griffith was one of seven engineers appointed by the Bog Commissioners to undertake surveys of the main bogland areas of Ireland. Between 1809 and 1813 he surveyed over 206,000 acres of bogland and made preliminary surveys of a further 267,000 acres of mountain bog. Griffith commenced his road and bridge building activities in 1822 when he was sent by the government to Munster to carry out a programme of public works as a means of alleviating famine and opening up parts of the counties of Cork, Kerry and Limerick.

In 1825, Griffith was appointed Director of the Boundary Survey, a land and property valuation survey that ran in parallel with the general mapping being carried out by the Ordnance Survey. He returned to reside permanently in Dublin in 1828 at 2 Fitzwilliam Place. His later career embraced many public duties, including acting as a commissioner enquiring into the development of Ireland's railway system, and as a commissioner making recommendations for the improvement of the Shannon Navigation. He was Chairman of the Irish Board of Works 1850-1864.

Griffith was a founder member of the ICEI and was twice its president: during the period 1850-1855 and again in 1861. He was elected to membership of the RIA in 1819. He received an honorary degree of LLD from the University of Dublin in 1849 and an honorary master of engineering (MAI) degree from the same university in 1862. He was created a baronet on 8 March 1858. Richard Griffith died at 2 Fitzwilliam Place in Dublin on 22 September 1878 and was buried in Mount Jerome Cemetery.

HEMANS, George Willoughby, MRIA (1814-1885), civil and railway engineer, was born at "Bronwylfa", Ryllon, near St Asaph, N.Wales on 27 August 1814, the eldest son of Capt. Alfred Hemans of the 4th (King's Own) Regiment and his wife, the poetess, Felicia Dorothea née Browne (1794-1835), who were married in 1812. There is a memorial window to his mother in St Anns Church in Dublin. In 1818, following a short and unhappy marriage, his parents separated and his father, who had long been in poor health, moved to Rome.

George's early education was undertaken by his mother, who remained at St Asaph. Hemans then spent three years at the Military College at the Abbaye École de Soréze in southwest France, where he won every prize in foreign languages, science and drawing, and left with no less than six medals.

As his mother had died in 1835 after a long illness, Hemans career was guided by his uncle, Colonel Browne, a Dublin magistrate. Browne obtained some work for Hemans with the Ordnance Survey in Ireland following which he was able to place his nephew as a pupil with John (later Sir John) Macneill at his office in London. This was probably the most important step in the young Heman's career as Macneill employed him on surveys of several lines of projected railway, both in Scotland and in Ireland.

Around 1840, Macneill appointed Hemans as Resident Engineer on the Dublin & Drogheda Railway (D&DR) and entrusted to him the complete works, including the erection of Ireland's first major wrought-iron lattice girder railway bridge over the Royal Canal in Dublin. For his paper to the Institution of Civil Engineers on the subject of the bridge, Hemans received a Walker Premium. Following the opening of the D&DR in 1844, Hemans worked for Macneill on the Dublin & Carlow railway, including part of the Great Southern & Western Railway (GS&WR) main line between Dublin and Cork. His address as a member of the Institution of Civil Engineers of Ireland was given as Rutland Square, the location of Macneill's Dublin office.

An Act incorporating The Midland Great Western Railway Company of Ireland (MGWR) had received Royal Assent in July 1845 and gave the new company powers to construct a railway from Dublin as far as Mullingar and Longford and also to purchase the Royal Canal company. Sir John Macneill (1793-1880) had been appointed Consulting Engineer to the company the previous year, but in 1845 resigned to take up a similar position with the competing Irish Great Western Railway (IGWR), which was being promoted by the Great Southern & Western Railway, of which Macneill was also Engineer. The IGWR planned to reach the Shannon at Athlone via Tullamore by a branch off the main Dublin-Cork line at Portarlington, a route that is currently used by Inter City services from Dublin to Galway. Following the resignation of Macneill, Jonas S Stawell was instructed to proceed with surveys from Mullingar to Galway, and his route was accepted on 16 October 1845 over that proposed by Macneill. Hemans was appointed on 21 August 1845 as 'Acting Engineer of the MGWR and branches'. It is likely that he became Chief Engineer some time after Stawell's resignation on 19 June 1846.

Now an experienced young engineer, Hemans was responsible for the design and supervision of construction of the line from Dublin westwards to Mullingar. The section from Enfield to Mullingar caused considerable difficulties, constructed as it was across deep bog in order to avoid the severe curves of the Royal Canal. Hemans overcame the problems with innovative engineering, literally laying the foundations for the design of both road and rail transportation routes across the bogs of the midlands and west of Ireland. The main line from Dublin to Galway included major bridge crossings of Lough Atalia and the rivers Shannon and Suck. The line to Galway was opened on 20 July 1851. Hemans was connected with several other railway companies, including lines in Ulster and Munster. He is said to have constructed more railways in Ireland than any other engineer of his time.

In 1854, Hemans moved to London and rapidly attained a deserved reputation as a parliamentary engineer. He lived with his wife at 13 Queen Square in Westminster and later at 32 Leinster Gardens, Bayswater. He maintained an office in Dublin at 46 Upper Sackville Street and in London at 26 Duke Street in Westminster. Railways in England and Wales, constructed under his supervision, included the Vale of Clwyd, and lines in Sussex and Herefordshire. In East Sussex, he lodged plans for lines from East Grinstead to Tunbridge Wells via Groombridge (1861), from Hartfield to Uckfield (1862). Plans were also lodged for the Daventry Railway in Northamptonshire with branches to Southam and Leamington (1863), whilst plans for the Tewkesbury & Malvern Railway were lodged in 1863, including a major bridge spanning the river Severn. In the Vale of Clwyd Hemans acted as Consulting Engineer for the line from Rhyl to Denbigh and on to Corwen in North Wales. He also engineered the lines from Zurich to Chur and Rheineck to Chur along the Upper Rhine valley in Switzerland for the Eastern Railway. John Mortimer Heppel (1817-1872) took charge of the Eastern Railway for Hemans before taking over in 1857 from George Barclay Bruce (1821-1908) as Chief Engineer of Madras Railways in India. Bruce became a London consultant and later, in 1872, helped Hemans to continue his work for the New Zealand government following the latter's stroke.

Jointly with the leading civil engineer of the time, John Frederic La Trobe-Bateman (1810-1889, Hemans prepared plans for the utilization of London's sewage. These works were actually commenced, but due to the financial crash of 1866 and the ensuing depression, the project was abandoned. In the same year, with Richard Hassard, he published *'On the future water supply of London.'* The proposed scheme would have supplied water to London from the Lake District via a lengthy aqueduct.

In 1861, Hemans laid plans for connecting the GS&WR at Kingsbridge (now Heuston) Station with the MGWR at Cabra. This necessitated a viaduct across the Liffey and a tunnel under the Phoenix Park. The link was completed by 1877. He also planned the line from Liffey Junction to the North Wall and Spencer Dock at the entrance to the Royal Canal. He advocated the construction of a new cattle market at North Wall, served by rail, to replace that at Smithfield, but the dealers objected to the proposed location and the cattle market was eventually built near Parkgate Street, cattle having to be driven through the streets to the quayside for onward shipment.

When it was planned to open it as a public park, Hemans presented a design for improvements to St Stephen's Green, including a grand entrance opposite the top of Dawson Street, but his design was not adopted. The same year (1864) he designed entraining embankments in County Clare to reclaim much of the slob lands at the mouth of the River Fergus.

By the 1860s, the growing problem of disposal of sewage was exercising the minds of the leading engineers. Hemans proposed spreading Dublin's sewage over the North Bull, but the residents of Clontarf objected. By 1870, Hemans, who had been engineer to a sewage utilisation scheme for London (not proceeded with), which would have entailed conveying the sewage by a 10ft diameter, 40-mile long culvert out to the Maplin Sands, was still advocating spreading sewage over land (a form of recycling) rather than dumping at sea.

Railway work having declined in Britain and Ireland, Hemans sought work abroad. In 1870 he was appointed Engineer-in-Chief for the Province of Canterbury in New Zealand, and subsequently Engineer-in-Chief to the New Zealand Government. However, in 1872, he suffered a severe stroke that left him paralysed and incapable of speech or writing for the rest of his life.

Hemans was a council member of the Institution of Civil Engineers of Ireland from 1849 and served as President in 1856-7, being the first to deliver a presidential address. He was also an active member of the Institution of Civil Engineers, being elected an associate member 2 May 1837 and transferring to membership 18 February 1845. He served on Council from 1856 and as a senior Vice-President until 1875, when he resigned due to failing health. He had been elected a Member of the Royal Irish Academy as early as June 1840, a sure sign of the confidence placed in him by his mentor, Sir John Macneill.

On 1 February, 1841, Hemans married Anne Cunison née Drysdale of Jerviston House, Bothwell, Lanarkshire, Scotland. They had three sons and three daughters. In 1866 his daughter Mary Harriet married the civil engineering contractor, Thomas Selby Tancred (1840-1910), who had trained under his father-in-law. Tancred joined with Hemans and a former colleague Travers Frederick Falkiner, to form the London consultancy of Hemans, Falkiner & Tancred with offices at 1 Westminster Chambers in Victoria Street.

George Willoughby Hemans died on 29 December 1885 at 11 Roland Gardens, Brompton, London.

MULLINS, Michael Bernard (1808-1871), civil engineer, was born in county Westmeath on 1 August 1808, the eldest son of Bernard Mullins, contractor, and Bridget Maria, daughter of Michael Hoey from county Westmeath. He entered Trinity College Dublin in October 1821, graduating from the University of Dublin with BA (1826) and MA (1834).

He spent most of his life working in his father's contracting business, the Dublin offices of which were located at 9 Mabbot Street in Dublin. His father died in 1851.

With his father Bernard, who was a vice-president of the ICEI, he presented an important paper to the Institution in 1846 'On the origin and reclamation of peat bogs, with some observations on the construction of roads, railways, and canals in bogs'.

Mullins's presidential address formed a detailed historical account of Irish civil engineering up to around 1860 and ran to 186 pages of the *Transactions of the Institution of Civil Engineers of Ireland*, and was delivered over three evenings.

Mullins country seat was at Ballyegan in county Offaly. Michael became a JP for the county and a high sheriff in 1838. In 1861, he was appointed a director of the Great Southern & Western Railway. His Dublin residences were at 1 Fitzwilliam Square South from 1834 and later at 18 Fitzwilliam Square.

Mullins was elected MICEI in 1847, served on the council 1852-55, was vice-president 1856-7, 1863-65, and served as president 1860-61.

Mullins died unmarried at 18 Fitzwilliam Square in 1871. In his will he left the bulk of his estate for the establishment of St Vincent's Convalescent Hospital in Grove Avenue, Blackrock. He bequeathed the contents of his office - drawings, books, models and drawing instruments - to the Institution of Civil Engineers of Ireland. When the Institution moved from Dawson Street in the early 1960s the loose architectural drawings from the collection were misappropriated and subsequently sold at auction by Allen & Townsend in 1965; these are now dispersed in various places, including the National Library of Ireland, the Irish Architectural Archive, Castletown House and Tullynally, county Westmeath. The Institution retained a large bound album of drawings from his office, which is now deposited in the Irish Architectural Archive at 45 Merrion Square in Dublin.

VIGNOLES, Charles Blacker, (1793-1875), civil engineer, was born at Woodbrook near Enniscorthy in county Wexford on 31 May 1793. Descended from a notable French Hugenot family, his father, Captain Charles Henry Vignoles, was stationed in Ireland in 1793 and afterwards in the West Indies.

Following the death from yellow fever of both his parents in Guadaloupe, Vignoles became a prisoner-of-war. Although still an infant, he was given a commission in the army as a means of effecting his release. He was returned to England and placed under the care of his maternal grandfather, Charles Hutton, FRS, a mathematical professor at the Royal Military Academy at Woolwich. Hutton 'adopted' Vignoles as his own and laid the foundations of a sound and liberal education.

Around 1807, Vignoles was articled to a law firm, but at the age of 19, was sent to the Royal Military Academy at

Sandhurst under the care of Professor Leybourne. The following year, he saw active service at the Battle of Vittoria and later received a commission in the Royal Scots. After a spell of duty in Canada, Vignoles was sent to Fort William in Scotland and from there in May 1816 to Spain.

He married Mary Griffith in 1817, went on half pay from the army, and sailed alone for South Carolina, where he carried out surveys and produced an account of the Dominion of Florida, accompanied by a highly acclaimed map.

Vignoles returned to England in May 1823, and was engaged in the summer of 1825 by the Messrs Rennie on surveys in Surrey and Sussex of a line of projected railway from London to Brighton. Shortly afterwards, he undertook surveys for the proposed Liverpool and Manchester Railway, including the route over Chat Moss. He was appointed resident engineer by the Rennies on 12 August 1825 and, in 1826, drew up the first contracts and began work on draining Chat Moss.

He was appointed engineer for the St Helens and Runcorn Gap Railway in 1829, and a line from Parkside on the Liverpool and Manchester Railway to Wigan, the origin of the North Union Railway, with which Vignoles was to be involved for many years. At the same time, he was involved with proposals for a line between Barnsley and Goole, and the London and Birmingham Railway, eventually realised as the Grand Junction Railway. In 1829, he resumed his links with Ireland when, with John Collister's assistance, he surveyed improvements on the Slaney Navigation.

The subsequent professional career of Charles Blacker Vignoles saw many achievements, including being given credit for the invention of the eponymous 'Vignoles Rail', although it seems more likely that he introduced it to Europe from the USA. In 1832 he was appointed engineer to the Dublin & Kingstown Railway and was responsible for the design and supervision of the construction of this, the first passenger railway in Ireland. He was also responsible for the construction of numerous other railways in Europe and South America, and also for the Nicholas suspension bridge over the River Vistula at Kiev in the Ukraine. In 1836 he was an engineer appointed by the Irish Railway Commission to investigate railway routes in the south and southwest of Ireland. A detailed biography of Vignoles may be found in *Dictionary of Civil Engineers in Great Britain and Ireland, Volume 2: 1830-1890* published in 2008.

His first wife died in 1834 and he remarried in the Spring of 1849. He had five sons and two daughters by his first wife.

Vignoles was a founder member of the ICEI and served as president in 1863-4. He was elected a Fellow of the Royal Society in 1855 and was also a member of the RIA, elected 1836. He became president of the Institution of Civil Engineers in 1870-71, having been elected MICE in 1827. Vignoles died at Hythe in Hampshire on 17 November 1875 and was buried in Brompton cemetery.

MALLET, Robert (1810-1881), mechanical engineer and iron-founder, was born on 3 June 1810 at Ryder's Row, Dublin, the son of John Mallet, iron-founder, and Thomasina Mallet. Robert, received his early education at Bective House Seminary for Young Gentlemen in Dublin. Prior to entering Trinity College Dublin in 1826, Mallet went on a tour of the continent with his future brother-in-law, William Watson, a civil engineer. The young Mallet exhibited an enthusiasm for engineering and developed a good grasp of French and German.

Having read classics, science and mathematics at Trinity College Dublin, he received a BA from the University of Dublin in 1830. On leaving university, Mallet first went on an extended tour of engineering works on the continent, before joining his father's iron-founding business in 1831. In time, Robert, who became a partner in the firm in 1832, built the Victoria Foundry into one of the most important engineering works in Ireland, undertaking large contracts, including much of the ironwork required by the major railway companies.

As part of the improvement of the Shannon Navigation in the 1840s, all the major bridges spanning the river were either replaced or rebuilt. They included opening navigation spans, a number of which were designed and supplied by J & R Mallet. In 1848 the firm supplied the castings for and erected the first lighthouse tower on the Fastnet Rock (replaced by the present masonry tower, completed 1904).

In his foundry, Mallet researched extensively into the properties and strengths of materials, corrosion of iron, and problems associated with the cooling of large iron castings, being one of the few early nineteenth-century iron founders to attempt a scientific explanation of fracture in terms of the 'molecular structure' of metal. He began to communicate papers on diverse topics to the Royal Irish Academy, to which body he was elected a member in 1832, when only 22 years of age.

Mallet was elected an Associate of the ICE in 1839 'because of having engaged extensively in the manufacture of machinery and possessed of considerable scientific attainments'. He was transferred to MICE 14 June 1842. He was elected to membership of the Civil Engineers Society of Ireland (later the ICEI) in 1836. As Secretary of the ICEI, Mallet was the driving force behind its reorganisation in August 1844 and was responsible for the introduction of printed transactions and for drafting the By-Laws. Having served on the council and as a vice-president for a number of years, in 1866 he became president for a two-year term. Shortly before his death in 1881, he received honorary membership of the ICEI. He was elected MIMechE in 1867.

In 1840, Mallet commenced experiments with wrought-iron plates of square or rectangular form hammered so as to curve them in both transverse directions thereby obtaining maximum strength for minimum weight. In 1852, he patented his so-called 'buckled plate' and such plates were used extensively in a variety of building applications, in particular for bridge decking. Mallet also patented a suggested improvement to the system of atmospheric railways, whereby the vacuum would be created by the stationary steam engines on a continuous basis and stored in large tanks, thus making the power source instantly available and independent of the frequency of the trains.

Another patent taken out by him was for an improved design of locomotive turntable and for a system of transferring private coaches transversely from road to flat rail wagons.

In November, 1831 Mallet married Cordelia Watson and had three sons, John Trefusius (engineer), John William (chemist) and Frederick Richard (geologist), and three daughters, all of whom married. The family moved in to a large house, 'Delville' in Glasnevin, a northern suburb of Dublin, but following the death of his first wife in 1854, Robert moved to a smaller house in Kingstown (now Dun Laoghaire). This was more convenient for travelling to and fro from London where he had set up consultancy offices at 11 Bridge Street, Westminster and later at 7 Westminster Chambers in Victoria Street.

By 1860, engineering work in Ireland had become scarce. The main railway routes had substantially been completed and arterial drainage operations suspended due to lack of funds. Mallet found that costs had become too high to permit profitable competition with English and Scottish ironworks, whilst the state of Irish industry afforded little opportunity for engineering works. Mallet's firm also failed to gain the contract for the supply of the pipework and other ironwork for the Vartry Water Supply Scheme for Dublin (opened 1865). As a consequence, and with his father now in his 80s, it was decided to close the Victoria Foundry. Mallet then moved permanently to London, purchased a house in Clapham, and in 1861 married Mary née Daniel, the daughter of the owner of the house where he had lodged whilst looking for a suitable residence. He had been spending increasing amounts of time in London looking after his growing consultancy and attending to his scientific writings. Mallet invested much of his time in the 1850s in attempting to solve the problems involved in constructing large calibre siege mortars and is credited in 1855 with stating the principle of ringed ordnance. The principle involved increasing the bursting strength of the mortar by building up the whole thickness using superimposed lamina or rings with initial tension, rather similar to present-day pre-tensioning of structural components.

Mallet is best remembered for his investigations in physical geology, these being directed towards four main areas: glacial flowage (1837-45), geological dynamics (1835 onwards), seismology (1845 onwards), and vulcanology (1862 onwards). Mallet is credited with having coined no less than eight terms using the prefix 'seism-', including the word seismology. His classic paper to the Royal Irish Academy in 1846 on earthquake dynamics is regarded as one of the foundations of modern seismology and in 1862 he was awarded the Cunningham Medal by the RIA for his researches in this area. In the same year, he was the recipient of an Honorary Master of Engineering from the University of Dublin and two years later received an Honorary LL.D. He was elected FRS in 1854.

Around 1870, Mallet began to lose his eyesight and used a secretarial assistant to continue his technical writing. Robert Mallet died on 5 November 1881 at 'Enmore', 2 The Grove, Clapham, London of chronic cystitis and is buried in Norwood Cemetery.

ANDERSON, Sir William, (1835-1898), mechanical engineer, was born in St Petersburg on 5 January 1835, the fourth son of John Anderson, a merchant banker, and Frances Simpson. William was educated at St Petersburg High Commercial School, where he was head of school and a silver medallist. Although a British subject, he was conferred with the freedom of the city. In 1849 he became a student on the three-year Applied Sciences course at King's College, London, graduating in 1852 as an Associate. On leaving college, he served a three-year pupilage at the works of Sir William Fairbairn at the Canal Street Works in Manchester and was engaged on the erection of machinery in Wales and in Ireland, where Fairbairn's contracts during 1852 included a steam-powered iron scoop wheel for the Wexford Harbour reclamation works, a large cast-iron suspension waterwheel for the Midleton distillery in county Cork, and the railway viaduct over the river Suir at Cahir in county Tipperary.

In 1855 he joined the engineering firm of Courtney & Stephens at the Phoenix Foundry in Blackhall Place in Dublin, eventually becoming a managing partner. He worked on railway signalling apparatus and the design of cranes. He gave considerable thought to the theory of diagonally braced girders and is credited with the introduction of the braced web in bent cranes. He was also involved with the erection of iron bridges, including the Malahide Viaduct on the Great Northern Railway (Ireland), completed in 1861. During this time, Courtney & Stephens supplied a large wrought-iron scoop wheel, believed to be the largest in existence at the time of installation, for the southern reclamation area at Wexford Harbour.

During his time in Dublin, Anderson met with an accident when one of his arms was caught in machinery and the elbow seriously injured. Amputation of the arm was avoided, but he became generally left-handed.

In 1864, Anderson moved back to England and joined the firm of Easton & Amos of Southwark. His subsequent career in England and abroad may be followed in his biography in the *Oxford Dictionary of National Biography.* On 11 August 1889, Anderson was appointed Director-General of the Royal Ordnance Factories at Woolwich Arsenal and elsewhere.

Anderson was elected a member of the ICEI 27 May 1856, served on the council and contributed important papers to the transactions. He became president in 1867. Elected MICE 12 January 1869, he became a member of council in 1886, and in 1896 was elected a vice-president. A regular contributor to the *Proceedings of the Institution of Civil Engineers*, he was awarded a Watt Medal (1872) and two Telford Premiums for presented papers. Being fluent in Russian, Anderson translated many abstracts of foreign papers for inclusion in the proceedings, including Chernoff's researches on steel. He was president of the IMechE in 1892/3, was elected FRS 4 June 1891. He received a C.B. 28 June 1895 and a Knighthood 1 January 1897. He was conferred with an Honorary DCL by the University of Durham in 1889 when he was president of Section G of the British Association meeting at Newcastle.

On 11 November 1856, Anderson married Emma Eliza, daughter of the Rev J.R.Brown, Incumbent of Knighton, Radnorshire. The couple had at least two sons, Edward William and Kenneth.

Sir William Anderson died on 11 December 1898 at his official residence, Woolwich Arsenal, in London following a heart operation.

GREENE, John Ball (1821-1896), surveyor, was born in Dublin in 1821, the third son of George Greene of Dublin and Jane Ball of Ball's Grove, Drogheda.

It is most likely that he received his training in the offices of Brassington & Gale in Bachelor's Walk in Dublin. He was occupied as a surveyor and valuator of estates, and is listed as living at 5 George's Place North in the north city. Following a period working under Isambard Kingdom Brunel on the Great Western Railway in England, by 1853 Greene had joined the staff of the General Valuation and Boundary Survey of Ireland as General Superintendent under Richard Griffith. In 1868 he took over Griffith's valuation duties and became Commissioner of Valuation, while Griffith retained the position of General Boundary Surveyor until at least 1875. On, or perhaps before, Griffith's death in 1878, Greene succeeded him in the latter position.

He was knighted in 1885 and retired in 1892. He was also a magistrate for county Dublin. Greene was married twice: firstly, in 1850, to Ellen, daughter of Robert Wesley, a surgeon in the Navy; secondly, in 1867, to Charlotte Mary, daughter of Edward Henry Courtenay of Cheltenham. A son, G C Ball Greene was present at the death of his father. His brother, Henry, was one of the contractors on the Vartry water supply tunnel. He lost everything as a result of the difficult rock conditions and died tragically as the result of a fall from a wall at Glendalough. He was only 45, and left a wife and fourteen children destitute.

A generously subscribed endowment fund rescued the family from immediate distress.

John Ball Greene was elected MICEI in 1857, served as a council member in 1864, 1876, and again in 1885. He was vice-president in 1867 and president in 1869-1871. He died at 53 Raglan Road, Dublin on 4 February 1896.

STONEY, Bindon Blood, (1828-1909), civil engineer, was born at Oakley Park, county Offaly (formerly King's County) on 13 June 1828, the second son of George Stoney and Anne, daughter of Bindon Blood of Cranagher and Rockforest, county Clare. His elder brother was George Johnstone Stoney (1826-1911), the distinguished scientist who coined the word 'electron'.

In 1845, Bindon Blood Stoney, having been educated privately, entered Trinity College Dublin. Following a thorough grounding in the classics leading to the award of his BA degree in 1850 by the University of Dublin, Stoney, having studied civil engineering, received a Diploma in Civil Engineering in the same year, with distinction in a number of subjects. In 1870, he received the degrees of MA and MAI from his alma mater, which, in 1881, conferred on him an Honorary Degree of Letters (LL.D).

On leaving university, Stoney became an assistant to the Earl of Rosse in the Observatory at Parsonstown (now Birr) where he ascertained the spiral nature of the Andromeda nebula. using the 72in reflecting telescope, the largest in the world at the time. His first professional engineering engagement was as an assistant engineer working for William Greene on surveys for the Aranjuez to Alamansa railway south of Madrid. Returning to Ireland in 1853, he became resident engineer under James Barton on the Boyne Viaduct. The central bridge section of the 1760 ft long viaduct comprised three spans of lattice girders forming the sides, bottom and top of a box, the central span of 264 ft being probably the longest of its type in the world at the time of its erection in 1855. The bridge was the first of its kind in which, upon a large scale, the strength of each part was accurately proportioned to the load it had to withstand. This saved material and, by reducing the weight, resulted in a more efficient design. Stoney's connection with its construction encouraged him in later years to write his classical work on *'The Theory of Strains in Girders and Similar Structures'*, first published in two volumes in 1866/69.

Following the completion of the Boyne Viaduct in 1855, Stoney spent some time on drainage works in the West of Ireland before being appointed to work under George Halpin Junior (Inspector-of-Works) at

Dublin Port. In 1859, Stoney was appointed Executive Engineer and in 1862 succeeded Halpin. The reorganization of the Ballast Board in 1867 and the establishment of the Dublin Port & Docks Board resulted in Stoney being confirmed as Chief Engineer at Dublin Port at a salary of £1000 per annum, an appointment that he held until his retirement in 1898. His salary was increased in 1866 to £1500 and in 1884, in recognition of his contribution to the development of Dublin Port, to £2000.

When Stoney entered the service of Dublin Port, its condition was that of a tidal harbour with a shallow approach channel from the Dublin Bay to the city. He designed large dredging plant, including hopper barges of 1,000-ton capacity for conveying the dredged material out to sea for dumping in deep water. The economies thus effected allowed the Board to press forward with further improvements to the approach channel to the port, so that, by the end of the nineteenth century, it was open at all states of the tide to vessels engaged in the cross-channel and coasting trade.

When Stoney began work at Dublin Port, there were no quays at which vessels could lie afloat at all states of the tide, except in the Custom House Docks, which belonged to the government, and the private docks of the Grand Canal and Royal Canal companies. Whilst Chief Engineer, he rebuilt the quay walls along both north and south banks of the river Liffey, representing half the shipping quays of the port, which replaced the tidal berths by deepwater berths at which large ocean-going vessels could lie constantly afloat. In addition to this, the northern quays were extended eastward and the formation of Alexandra Basin commenced.

It was probably in the construction of these latter quays that his ingenuity and resource were more especially recognized by the general public, but it was his construction methods that brought him to the attention of the engineering profession. To avoid the necessity of costly cofferdams and pumping, he formed the lower sections of the deep-water quay walls of monolithic concrete masonry blocks of up to 350 tons weight. These were built in a casting yard, and when sufficiently hardened lifted by a floating crane, transported to the site of the new quay wall, and lowered on to prepared foundations. The machinery for handling the blocks, and the large diving bell for preparing the foundations, were all designed by Stoney, who described the work in a paper delivered at the ICE in 1874, and for which he was awarded a Telford Medal and Premium.

In addition to harbour works, Stoney directed the design and rebuilding of two major bridges across the river Liffey, namely Grattan Bridge and O'Connell Bridge. Grattan (previously Essex) Bridge was erected in 1676 and rebuilt in 1753-55 by George Semple with semi-circular arches and steep approaches. The present bridge, designed by Stoney, has flatter elliptical arches, a level roadway and footpaths cantilevered on wrought-iron brackets beyond the face of the arches. O'Connell (previously Carlisle) Bridge, on the line of the main artery through the city and erected in 1791, was also too narrow and hump-backed for modern traffic and its replacement was opened in 1880. The new bridge, built in granite ashlar, is as wide as the thoroughfare that it connects and has elliptical arches.

On 7 October 1879, Stoney married Susannah Frances, daughter of John Francis Walker, a barrister, and had four children, including a son George.

Stoney was elected a member of the Royal Irish Academy in 1857 and was also a fellow of the Institution of Naval Architects. He was elected a fellow of the Royal Society in 1904. He was elected MICEI in 1857, served on the council, and was honorary secretary between 1862 and 1870, before being elected president in 1870. He was elected an associate of the ICE in 1858, and was transferred to MICE in 1863. He also served on the council of the ICE for a number of years.

Bindon Blood Stoney died on 5 May 1909 at 14 Elgin Road, Dublin, of chronic bronchitis, and is buried in Mount Jerome cemetery.

COTTON, Charles Philip (1832-1904), railway engineer, was born in Dublin on 19 January 1832, the son of The Venerable Henry Cotton, Archdeacon of Cashel and his wife, Mary Vaughan, youngest daughter of Richard Laurence, Archbishop of Cashel.

Cotton received a classical education at St Columba's College, Stackallan, county Meath. In 1848, he entered Trinity College Dublin and, having obtained a Junior Moderatorship in Experimental and Natural Sciences, studied civil engineering, obtaining his Diploma in Civil Engineering in 1854 with two certificates of distinction in practical engineering and in chemistry and geology. He had been awarded his BA in the previous year.

He then served a pupilage under William Richard LeFanu for whom he acted as resident engineer for two years on the construction of the Bagenalstown & Wexford railway, and for a similar period on the Mallow & Fermoy railway, becoming in time chief assistant to Le Fanu.

Following the retirement of Le Fanu in 1863, Cotton acted as Engineer-in-Chief and completed the line from Wicklow to Wexford. This included two multi-span masonry arch viaducts at Rathdrum over the Avonmore River and short lengths of tunnel at Rathdrum and Enniscorthy. He also completed the Shillelagh Branch from the main line at Woodenbridge and the Roscrea & Nenagh Railway. In 1877 Cotton completed the extension of the Dublin South Eastern Railway from Macmine Junction via New Ross to Waterford, the route necessitating some heavy earthworks, a tunnel and a five-span opening bridge crossing of the river Nore. Cotton was also consulted by the Cork & Bandon Railway when it became necessary to replace a timber lattice bridge at Innoshannon over the Bandon river with an iron lattice girder bridge. The bridge was designed by Le Fanu, the ironwork being supplied by the Cork Steam Ship Company. During the period 1866 to 1878, Cotton was in partnership with Benjamin Flemyng, with offices, first at 36 Westland Row, from 1872 at 3 Harcourt Place, and from 1876 at 136 St Stephens Green West, all in Dublin. In 1867, Cotton drew up plans for a new harbour at Bray in county Wicklow.

From 1863 to 1867 Cotton acted as Engineer to the 'Ballast Board', which then had charge of lighthouses, and was involved in the strengthening of the original Fastnet Rock cast-iron lighthouse. He also acted as county surveyor for Kildare between August and October, 1865.

In 1879 Cotton was appointed Chief Engineering Inspector of the Local Government Board, Ireland, which appointment he held until his retirement in 1899. In 1898, he was appointed a member of the Royal Commission on Sewage Disposal and, in 1900, Chairman of the Vice-Regal Commission to inquire and report on the health of the City of Dublin.

He was author of a number of books, including *A Manual of Railway Engineering in Ireland* (1861) and *The Irish Public Health Acts 1878-1890* (1891)

In 1878 Cotton married Marion Louisa, eldest daughter of Sir Maltby Crofton, Longford House, county Sligo. They had no children.

Cotton was elected AMICE in 1861 and transferred to MICE in 1864. In 1861, he was elected MICEI and became president in 1874. He was elected MRIA in 1864.

Cotton died at "Ryecroft", Bray, county Wicklow on 10 March 1904 of heart problems leading to coma.

MCDONNELL, Alexander (1829-1904), locomotive engineer, was born in county Dublin on 18 December 1829, the third son of John McDonnell, a medical doctor and Poor Law Commissioner of Ireland, who practised at 32 Upper Fitzwilliam Street, Dublin.

Alexander entered Trinity College Dublin in July 1847 to study arts, but also attended engineering lectures in the newly founded School of Engineering, which offered a diploma in civil engineering at the end of two years study. He gained honours in mathematics but did not take an engineering diploma. He graduated BA from the University of Dublin in 1852 and was awarded his MA in 1861. He served a three-year pupilage with Charles Liddell of the firm of Liddell & Gordon, Westminster, and remained with them until 1858. During this period he spent some time at the École des Ponts et Chausées in Paris. He was resident engineer on the Newport, Abergavenny & Hereford Railway, based in Hereford, and worked on the Crumlin Viaduct. He also served as superintendent of the locomotive and engineering department of the company and remained in that position until 1861. He then undertook railway work in Turkey for two years before returning to England.

In 1864, McDonnell was appointed locomotive superintendent of the Great Southern & Western Railway (GS&WR) at Inchicore, where he was to spend the next eighteen years. During his time at Inchicore, the locomotive, carriage and wagon shops were greatly developed. McDonnell was responsible for the design of a range of locomotives, including types with 0-6-0, 2-4-0 and 0-4-4 wheel arrangements.

In 1882, McDonnell moved to the North Eastern Railway until his resignation from the company in 1884. He then became an overseas consultant. He was elected MICEI in 1860, MICE in 1872, and became president of the ICEI in 1875.

On 29 July 1867, McDonnell married Arabella Blanche, daughter of George Grenfell, at the British Embassy in Paris. At the time of the 1901 census, he is recorded as living with his wife and unmarried daughter Maria at 43 Rydens Road in Walton-on-Thames, Surrey. He died, whilst returning on a visit to Ireland, at The Station Hotel, Holyhead on 4 December 1904. The funeral took place on 8 December 1904 to Kilsharvan near Drogheda.

MANNING, Robert (1816-1897), civil and estate engineer, was born in Avincourt, Normandy, France on 22 October 1816, the third son of William Manning of Knocknamohill near Arklow in county Wicklow and Ruth, daughter of Lionel Stephens of Dromina in the parish of Crooke near Passage East in county Waterford. William, a commissioned army officer was on garrison duty in Normandy when his son Robert was born. Robert was educated privately at Kilkenny and at Waterford. He did not receive any education or formal training in fluid mechanics or engineering, but learnt accounting..

Following the appointment of Samuel Ussher Roberts as District Engineer for the Glyde and Dee catchments in county Louth, he recruited Manning in April 1846 to act as Accountant and Draughtsman in the District Office. Nine months later Manning was promoted to Assistant Engineer. In this capacity, he not only assisted Roberts in the drainage works on the east coast, but was also responsible for surveying and reporting on the design of river improvement works in the counties of Sligo and Donegal in northwest Ireland. In January 1848, Roberts was transferred to county Galway and

Manning was appointed as District Engineer on the Ardee and Glyde works. As District Engineer, the preservation of working water-powered mills on the rivers undergoing drainage work was one of the recurrent difficulties he had to face.

By 1854, the drainage works were being closed down and Manning was forced to seek employment elsewhere. He was appointed in 1856 to carry out a topographical survey of the Irish estates of the Marquis of Downshire covering over 120,000 statute acres and situated in five different counties. In 1858, on completion of the survey, Manning was appointed Estate Engineer to the Marquis, a position that he held for ten years until the post was abolished by a subsequent Marquis.

During the 1860s, Manning carried out extensive scientific studies on aspects of rainfall, river volumes and water runoff in respect of the mooted use of rivers on the Downshire Estate for a water supply for Belfast. He ventured into new areas of theory and practice that led to a paper delivered to the ICE in May 1866 for which he was awarded a Telford Gold Medal and a Manby Premium.

In December 1868, Manning was again looking for work. In October 1869 he was appointed as Second Engineer in the Board of Works and on 1 April 1874 was appointed Chief Engineer on a fixed salary of £800 per annum and served in that capacity until 1891 when he retired at the age of 75. During this time he was in charge of all engineering works of the department, including work at the five Royal harbours, nearly 200 fishery piers and harbours, of which he designed and constructed upwards of 100, and improvement works on the river Shannon (commenced 1880). Despite his heavy workload, Manning found spare time at home to study the classical nineteenth-century experiments on open channel flow in the laboratory and in the field. This resulted in his classic paper in 1889 entitled 'On the flow of water in open channels and pipes'. This was followed by a further paper on the subject in 1895.

Manning pointed out the confusion at the time surrounding even the rough calculation of 'the velocity or surface inclination of water', noting that there was no such thing as true uniform motion in river channels and that 'even to observe and record correctly the physical data required was a matter of extreme difficulty', but that nevertheless something better than an 'empirical formula' or 'rule of thumb' method ought still to be possible. He put forward a formula, checking its effectiveness against some 413 experimental measurements of open channel flow and 230 measurements of pipe flow. Within a few years his monomial formula for open channel flow passed into the standard literature and became one of the building blocks of the modern science of hydrology. Today it is still the most widely used formula for open channel flow.

Manning was elected AMICEI in 1848, transferred to MICEI in 1856, served on council, and was president in 1877-1878 when the Institution received its Royal Charter. Manning was elected MICE in 1858.

On 7 March 1848, Manning married Susanne Gibson of Baggot Street, Dublin. They had four sons and four daughters. One of his sons, William (1848-1903), became a civil engineer and another, Robert (1860-1894), worked as a surveyor on Canadian Railways.

Robert Manning died on 9 December 1897 at his residence 4 Upper Ely Place, Dublin of a heart attack and is buried in Mount Jerome cemetery.

BAILEY, John (1839-1892), mechanical engineer, was born at Wallop, near Stockbridge, Hampshire on 31 May 1839, the fourth son of Hinton Richard Bailey of Pittleworth Farm House, Hampshire and his wife Charlotte. Bailey received his early education at a private school in Southampton.

In 1857 he commenced a five-year apprenticeship with Summers and Day, engineers and shipbuilders, Northam Ironworks, Southampton. On the completion of his apprenticeship he was employed from 1862 to 1864 in the drawing offices of H.M.Dockyards at Devonport and Portsmouth.

In 1865, Bailey joined the Dublin engineering firm of Courtney and Stephens in Blackhall Place in place of William Anderson, the firm then becoming known as Courtney, Stephens, and Bailey, Bailey acting as chief engineering partner. During his nineteen years with the firm, important railway works were undertaken, including bridges and other railway infrastructure for the Midland Great Western, Dublin, Wicklow & Wexford, Great Southern & Western, and Great Northern railway companies. Between 1876 and 1884, Bailey carried out a number of major contracts for William Hemingway Mills, the chief engineer of the Great Northern Railway. The firm also provided the ironwork for the rebuilding of Essex Bridge in Dublin, the new roofs of Westland Row station, and the Nore viaduct at Thomastown in county Kilkenny. He often appeared as an expert witness in engineering lawsuits.

In 1866, Bailey married Jane, daughter of William A Summers, a Hull-born engine-maker of Southampton, with whom he had earlier served his apprenticeship.

He was elected MICEI in 1865 and served as president 1879-1880. In 1870 he was elected MICE. Bailey retired in 1884 and died at Brighton on 5 August 1892 from peritonitis.

NEVILLE, Parke (1812 - 1886), civil engineer, was born in Dublin in 1812, the eldest son of Arthur Neville, surveyor to the Corporation of Dublin. Parke Neville was first articled to Charles Vignoles and was engaged by him on the Dublin & Kingstown Railway (opened 1834) and for a brief period in England. Leaving Vignoles, Neville served a further pupilage with William Farrell, Architect to the Ecclesiastical Commissioners for Ireland. From around 1840, Neville was in private practice as both engineer and architect, carrying out many public works, including prisons, asylums and churches.

In 1845 he was appointed to act as Joint City Surveyor (probably with Arthur Neville and Charles Tarrant), all of whom did not receive a regular salary, but were paid for any work carried out. Following the reconstitution of the Corporation, Neville was appointed in 1851 as Borough Surveyor (later City Engineer) of Dublin and immediately set about improving the surfacing of the streets, and planning for a high pressure water supply and for the main drainage of the city. He lived at 79 Lower Leeson Street.

He drew up plans for the main drainage of Dublin as early as 1853, which included north and south interceptors and tidal discharge into the sea. In 1869/70, Neville, in conjunction with Sir Joseph Bazalgette, prepared detailed plans for a main drainage scheme and obtained an Act of Parliament in 1871. The design for high- and low-level sewers north and south of the River Liffey was similar in many respects to that previously suggested by Neville. However, due to the cost involved, the city authority did not feel able to proceed with the project and the Act was allowed to lapse. The main drainage scheme was subsequently completed in 1906 by Neville's successor Spencer Harty.

Parke Neville's most successful work was undoubtedly the Dublin Corporation Waterworks (the Vartry Scheme). Despite much opposition, Neville showed great determination in promoting the scheme with Sir John Gray, and had the satisfaction of the seeing the scheme completed in 1865 at a cost of £650,000. Much of the detailed design was undertaken by Neville, although the scheme had been put forward to the Dublin Pipe Commissioners as early as 1853 by Richard Hassard. The River Vartry in county Wicklow is impounded by an earth dam near Roundwood, water from the reservoir so formed being led to treatment works and thence by aqueduct to service reservoirs near Dublin for subsequent distribution. The scheme was expanded in later years by the completion of a second reservoir and a doubling of the capacity of the aqueduct.

In 1836 Neville married Eliza Waring (1807-1893) by licence and had a son, Parke Percy (b.1850), and a daughter Anna (1842-1895).

Neville was elected MICEI in 1845, served on council, and was president in 1881. He was elected MICE in 1865, and was also a vice-president of the Institute of Architects.

Following visits to a number of towns in England to view public baths, and to attend the British Association meeting, he returned to Ireland where he died shortly afterwards on 13 October 1886, at 58 Pembroke Road, Dublin, of congestion of the liver. He is buried at Mount Jerome cemetery.

MILLS, William Hemingway (1834-1918), railway engineer, was born in Bradford, Yorkshire on 5 August 1834, son of Charles Mills, a railway guard, and his wife Hannah. Mills received a public school education before being articled to William Henry Barlow under whom for the next two years he had charge of construction works on the Midland Railway. He then acted as an assistant engineer under Peter Barlow and afterwards for two years as resident engineer for James Samuel on the construction of the Morayshire Railway in Scotland, including the viaduct at Craigellachie over the River Spey, completed 1863. Mills had, in 1860, been appointed Engineer & General Manager of the Morayshire Railway.

In 1864, Mills accepted an appointment to a railway in Andalusia in Spain, and in 1870 became Engineer and General Manager of Mexican Railways, and was responsible for the single-track standard-gauge main line between the port of Vera Cruz and Mexico City (opened January 1873), a distance of some 300 miles and overcoming a difference in elevation of some 8,000 ft.

In 1876 Mills returned from Mexico to become the first Chief Engineer of the newly amalgamated Great Northern Railways (Ireland). Moving to their head office at Amiens Street (now Connolly) Station in Dublin, he inherited works of very variable standards, and in time established a distinctive GNR(I) style. Major work to improve the facilities at Amiens Street were undertaken around 1880 and this work was under his direction, he having an important role in the design of the new infrastructure. A new block of offices were erected following a competition won by Charles Lanyon. The station too was completely remodelled.

Mills introduced standardised track-bed, with both keys unusually to the inside to facilitate inspection. Between 1876 and 1911, 700 miles of track were re-laid, 220 houses, 38 goods sheds, and 17 additional stations were built and 48 rebuilt. The most important of the stations completed were those at Lisburn (1878), Coalisland, Stewartstown and Cookstown (all 1878-9), Sion Mills (1883), Newtownstewart, Strabane, Warrenpoint (1891) and Dundalk (1893). During his time with the GNR(I), Mills was responsible for the erection of 177 new bridges, many in steel, and 37 station footbridges. He replaced the last of Macneill's timber viaducts (at Rogerstown on the D&DR). Mills retired from the GNR(I) in 1909.

Whilst abroad, Mills was married, a son, Arthur Edwin being born in 1867, later becoming a civil and mechanical engineer.

Mills was elected MICE in 1862, and in 1877 was elected MICEI. He served on the council from 1878, and became president in 1883.

Mills died on 12 January 1918 at his residence "Nurney", Silchester Road, Glenageary, county Dublin, of pneumonia and heart failure.

ASPINALL, Sir John Audley Frederick (1851-1937), mechanical engineer, was born in Liverpool on 25 August 1851, the second of three sons of John Bridge Aspinall, QC, and Betha Wyatt Jee. He was educated at Beaumont College, Berkshire and in 1868 became a pupil at the Crewe locomotive works of the London & North Western Railway, first under John Ramsbottom and then under Francis William Webb. At the end of his pupilage he was appointed assistant manager of the Crewe steelworks and remained there for about three years.

In 1875 Aspinall moved to Ireland, where he had been appointed manager of the Great Southern & Western Railway at Inchicore in Dublin, where Alexander McDonnell was locomotive superintendent. He spent eight years as manager and a further three as locomotive superintendent in succession to McDonnell, during which time he introduced vacuum brakes

99

to his locomotives, a trend that was followed in Britain.

In 1886 he was appointed Chief Mechanical Engineer of the Lancashire & Yorkshire Railway (L&YR), where he was to establish his reputation. He was succeeded at Inchicore by Henry Alfred Ivatt. Aspinall became general manager of the L&YR company in 1899. One of his significant achievements was the electrification of the Liverpool & Stockport Railway, the first mainline electrification in the UK.

In 1902 he was appointed associate professor of railway engineering at the University of Liverpool, and from 1908 to 1915 acted as chairman of the faculty of engineering, assisting with the foundation of the chair of engineering. He received an honorary doctorate from the university in 1922.

In 1874, he married Gertrude, daughter of F.B.Schrader of Liverpool. They had a son and two daughters.

Aspinall was elected MICEI in 1877, served on the council, and became president in 1884-6. He was president of the IMechE in 1909, and president of the ICE in 1918. He was knighted in 1917 and in 1936 was awarded the IMechE James Watt Medal for outstanding contributions to the advance of mechanical engineering.

Sir John Aspinall died at Woking on 19 January 1937.

GRIFFITH, Sir John Purser (1848-1938), civil engineer, was born at Holyhead on 5 October 1848, the son of Rev.William Griffith, a Congregationalist minister, and Alicia Evans. From the age of 13, he was educated at Dr Bigg's school in Devizes, Wiltshire, and then at the Moravian Fulneck School in Yorkshire. He entered Trinity College Dublin in 1865 to study civil engineering and obtained his Licence in Civil Engineering (LCE) in 1868. He was to receive an honorary MAI from the University of Dublin in 1914.

On leaving college, he became a pupil of Bindon Blood Stoney at Dublin Port and then, from early 1870, worked as an assistant county surveyor with Antrim County Council on road projects. On 4 April 1871 Purser Griffith gained a permanent position as second assistant engineer to Stoney at Dublin port, a partnership that was to last for twenty-seven years.

Shortly after his appointment at Dublin port, Purser Griffith, on 8 November 1871 married Anna Benigna Fridlezius Purser, nine years his senior and the only daughter of his father's life-long friend John 'Tertius' Purser and Anna Benigna Fridlezius. The marriage took place in the Moravian Church in Dublin and they went to live at "Greenane" in Temple Road in Rathmines. Later, on the death of Anna's father in 1893, Griffith moved into Rathmines Castle, where he was to reside until his death many years later in 1938.

During his time at Dublin port (1871-1912) many important improvements were made and it is clear that Griffith contributed significantly to the successful outcome of the Alexandra basin project (opened 1885) and the rebuilding of the river quays, particularly on the north side between 1900 and 1907. This work, carried out behind coffer-dams, entailed the removal of the timber wharves in front of the original quay walls, and the provision of mass concrete foundations to support new masonry quay walls. He also developed a major programme of suction dredging of the shipping channel and berths, introduced in 1895 as a follow-up to the system of bucket dredging with which he had also been involved as Stoney's assistant, but he was constantly hampered by lack of investment by the port authorities.

In January 1899, Purser Griffith took over the reins as Engineer-in-Chief from Stoney, who had retired on health grounds. At this time, Belfast was more popular with ship-owners, because Dublin lacked mechanical cargo handling facilities. Griffith set about providing ten 10 ton-capacity portal cranes and, in 1905, a 100 ton-capacity crane. In 1912, Griffith designed and supervised the erection of twin 'Scherzer' type lifting bridges on the north quays over the entrance to the Royal Canal. It is possible that Griffith became acquainted with William Scherzer's patented design when visiting the Chicago exposition in 1893. Griffith became frustrated at the reluctance of the port authority to build on what he had achieved in providing deep-water facilities. He was totally opposed to its financial policies and, in December 1912, decided that his only course of action was to seek early retirement and fight the Dublin Port & Docks Board from within. He handed over to his son John William and was duly elected to the Board in January 1915, but the disagreements developed into a major controversy culminating in his resignation from the

Board in July 1916. His son resigned the following day, so ending the Griffith connection with Dublin port. John William, and another son, Frederic Purser Griffith, joined their father in 1917 in a small engineering consultancy, Sir John P.Griffith & Partners at 6 Dame Street in Dublin. The work of the office was never extensive, but became the base for the development of a number of projects of national importance, in particular the development of the country's waterpower and peat resources. In 1920, Trinity College Dublin introduced the subject of harbour engineering into the engineering curriculum and appointed Purser Griffith Honorary Professor of Harbour Engineering.

Purser Griffith had earned his knighthood in July 1911, not for his work at Dublin port, but for representing Ireland on the Royal Commission appointed in March 1906 'to enquire into and to report on the canals and navigations of the United Kingdom'.

Purser Griffith was elected MICEI in 1871, becoming president 1887-1888 at the age of 39. He was elected AMICE in 1877 and transferred to MICE in 1883, represented Ireland on council from 1901, vice-president in 1916, and served as president 1918-1919. In 1879, at the age of 31, he delivered an important paper on 'Improvements of the Bar of Dublin by Artificial Scour', for which he was awarded a Manby premium. He also delivered the James Forrest lecture on 24 October 1916.

In 1917 Griffith served as chairman of the Irish Peat Enquiry Committee, set up by the Fuel Research Board to consider the utilisation of Irish peat deposits. The subsequent reports of this and later committees were, however, pigeon-holed and Griffith, frustrated by the inertia of successive governments and the seeming total lack of appreciation of the value of Ireland's natural resources, purchased a number of bogs and, in 1924, established one of the earliest machine peat harvesting operations in Ireland.

Seeking an alternative to imported coal, attention was directed towards the utilisation of waterpower for the generation of electricity. Purser Griffith acted as chairman of a sub-committee of the 1918 Water Power Resources Committee to deal specifically with the situation in Ireland. This and later committees considered that waterpower development was possible. Griffith and others favoured developing the hydro-electric potential of rivers near to the demand, such as the Liffey near Dublin, but a proposal to harness the river Shannon at Ardnacrusha in county Clare was put forward to the government by a young Irish electrical engineer, Thomas McLoughlin, and the Shannon Scheme as it became known was built by the German firm of Siemens-Schukertwerke and commissioned in 1929.

Griffith was elected MRIA in 1919 and served as a commissioner of Irish Lights from 1913 to 1933. He was appointed a vice-president of the Royal Dublin Society (RDS) in 1922, and in the same year was elected to the Seanad (Senate) of the Irish Free State legislature, remaining a senator until 1936. He was awarded the prestigious Boyle medal by the RDS in 1931. In 1936, he received the honorary freedom of the city of Dublin.

In 1938, Griffith received an honorary DSc from the National University of Ireland, a short while before his ninetieth birthday. He died at his residence, Rathmines Castle in Dublin, on 21 October 1938.

HARTY, Emanuel Spencer (1838-1922), civil engineer, was born in Tralee, county Kerry in 1838, the son of George Harty. Spencer, who rarely used his first name, served a three-year pupilage to the Dublin land surveyor Joseph James Byrne and then spent a further two years as Byrne's assistant.

In 1860 he secured a temporary position with Dublin Corporation and two years later was appointed to the permanent staff. He was intimately connected with the Dublin Corporation waterworks at Roundwood (the Vartry scheme), first as engineer on the works during their construction in the 1860s and subsequently as engineer-in-charge of the reservoirs and other elements of the scheme.

On 14 March 1887 Harty was appointed Borough Surveyor and Waterworks Engineer in succession to Parke Neville, a position he continued to hold until 1910. On taking over from Neville, Harty was involved with the drafting of the Dublin Corporation

Main Drainage Act and worked with consultant George Chatterton on the new main drainage system for Dublin.

Following the prolonged drought of 1893, Harty urged the Corporation to construct a second reservoir in the Vartry catchment and plans were completed under his direction and work commenced prior to his retirement.

Harty's retirement early in 1910 was mainly on account of failing eyesight, but at about the same time he is reported to have suffered from a 'very severe and prolonged illness' from which he eventually recovered. He was granted a pension of £1,000 per annum and had been made a Freeman of the city of Dublin on 2 September 1907. He was succeeded as City Engineer and Borough Surveyor by John George O'Sullivan.

On 7 November 1863, at St Thomas's Church in Dublin, Harty married Frances, daughter of John Sumption. Of their thirteen children, a son George Spencer Harty (born March 1873) was appointed engineer to the cleansing department of Dublin Corporation in 1905. The family lived at 112 Brighton Road and later at 44 Cullenswood Avenue in Rathmines.

Harty was elected MICEI in 1875, was a member of council 1885-1889 and 1891-1922, and served as president 1889-90.

He died, after a lengthy retirement, at "Ranelagh", 76 Merrion Road, Dublin on 19 August 1922, predeceased by his wife who died in 1907.

PIGOT, Thomas Francis (c1836-1910), civil engineer, was born in Dublin about 1836, the youngest son of David Richard Pigot, Chief Baron of the Exchequer in Ireland (1846-1873), and Catherine, eldest daughter of Walter Page of Araglin Mills, county Cork.

Pigot entered Trinity College Dublin on 19 November 1856 as a 'socius commitatus' (paying double fees in return for certain privileges), but left in 1858 to study civil engineering at the École des Ponts et Chaussées in Paris in 1858. At the end of the three-year course he was placed second in the final examination for the 'elèves externes' (students other than government students selected by the École Polytechnique) and was duly awarded the school's diploma. He then spent eight months at Fairbairn's engineering works in Manchester, after which he worked on the Bristol and South Wales Union and Mid-Wales railways up until the end of 1863. He was then employed on the Royal Sardinian and La Vendeé railway, before returning to Wales to work on the Wrexham and Mold railway.

In 1866 he was appointed the first Professor of Descriptive Geometry, Mechanical Drawing and Surveying at the College of Science for Ireland, where he also ran a course in civil engineering and building construction. This course proved so successful that 'Engineering' was subsequently added to the title of his professorship. He was responsible for the creation of a School of Engineering and had a particular interest in inland navigation, to which he gave prominence in his presidential address to the Institution.

It would appear that he resided at his father's house at 15 Merrion Square East in Dublin. Pigot retired from the College of Science in 1892 due to failing eyesight and general ill-health caused by overwork. He was then appointed one of the two chief examiners in science under the Department of Science and Art, a post that he retained until 1901. At this time he was living with his widowed elder brother (a barrister) and his sister at 14 Fitzwilliam Place.

He was elected MICEI in 1874, served as a council member from 1876 until his death; was vice-president, 1889-1890 and president, 1891-1892. He was also a member of the IMechE, elected 1877. Pigot died, unmarried, at 14 Fitzwilliam Place, Dublin on 20 May 1910.

SMITH, John Chaloner (1827-1895), civil engineer, was born at St Stephen's Green, Dublin on 19 August 1827, the eldest son of John Smith, a proctor of the Irish Ecclesiastical Courts.

Chaloner Smith entered Trinity College Dublin to study civil engineering, but did not take the diploma. Instead, he left after successfully completing his second year examinations and took sufficient courses to be later awarded his BA degree in 1849. In 1846 he commenced a pupilage under George Willoughby Hemans.

In December 1853 Hemans appointed him resident engineer on the Waterford & Limerick Railway (W&LR) during its construction. Four years later he became Engineer of the Waterford & Kilkenny Railway, a post he held until 1861 when the line was amalgamated with the W&LR. He then entered into partnership with the contractor John Bagnall to construct the Borris & Ballywillan, the Clara & Streamstown, and the Roscrea & Birdhill lines.

In February 1868, Smith ceased contracting and became Engineer and later Chief Engineer of the Dublin, Wicklow & Wexford Railway, his connection with this railway continuing until within a year of his death in 1895. The New Ross extension line of eighteen miles involved deep rock cutting, tunnelling and a bridge with navigation span over the river Barrow. He was also responsible for doubling the line between Kingstown (Dun Laoghaire) and Dalkey and the diversion of the main line at Bray Head in county Wicklow. He undertook the substantial rebuilding of Westland Row (Pearse) station in Dublin and, with William Hemingway Mills, Engineer of the GNR(I), the viaduct and bridge over the river Liffey to connect with Amiens Street (Connolly) Station (the so-called Loop Line, completed in 1890) (This line had been proposed by Smith as early as 1872, and only continual pressure and lobbying resulted in him obtaining the necessary Act of Parliament in 1884).

On 3 June 1857, Smith married Eliza, daughter of Stephen Parker, at Monkstown parish church, county Dublin

Smith was elected MICE in 1862 and MICEI in 1871. He served as Hon.Secretary of the ICEI from 1873 until 1887, when he resigned due to failing eyesight, but later had a successful cataract operation. The ICEI made great progress during his tenure of office; its membership increased, a Royal Charter was obtained in 1877, and accommodation was acquired for a meeting hall, offices and library. A Smith Testimonial in his memory is awarded annually for papers of merit. He was elected a member of the Royal Irish Academy in 1868.

During the last few years of his life he devoted considerable time to a study of the financial relations between Great Britain and Ireland, and was a strong advocate for substantially reducing the railway rates in Ireland under a state guarantee for the existing proprietors and gave evidence before the Royal Commission on the subject.

Smith was an avid collector of British mezzo-tint portraits and in time his collection was sold at Sotherby's. He compiled a four-volume catalogue of such portraits that became the recognised reference for collectors.

Chaloner Smith died on 13 March, 1895 at The Vicarage, Clyde Road, Dublin, of angina pectoris.

PRICE, James (1831-1895), civil engineer, was born at Monkstown, county Dublin on 18 January 1831, the second son of James Price. He was educated at Trinity College Dublin and obtained his Diploma in Civil Engineering in 1850 and a BA from the University of Dublin the following year. On leaving college, Price obtained a position with the General Valuation of Ireland and subsequently was employed by James Barton as a general assistant. From 1855 to 1857, he was resident engineer under Barton on the construction of the Banbridge Junction Railway. He then acted for eighteen months as engineer on harbour improvements at Wicklow. In 1859-60 he was resident engineer on the construction of the Cootehill & Ballybay line, and was then engaged for two years as a resident engineer in charge of the permanent way and works of the Dublin & Belfast Junction Railway.

At the end of 1862, Price was appointed Engineer-in-Chief of the Midland Great Western Railway of Ireland (MGWR) and also had charge of the Royal Canal (purchased by the railway company in 1845). In 1873, Price designed and had built two floating swing bridges over the Spencer Dock on the Royal Canal in Dublin for the MGWR. He introduced the bascule type bridge to Ireland (across the river Shannon at Drumsna on the Sligo branch of the MGWR).

Price was one of three engineers selected by Dublin Corporation to report jointly on the purification of the river Liffey and reported in 1874. On leaving the MGWR in 1877, Price commenced private practice. Amongst the works with which the practice was associated were the Lough Erne drainage scheme (1882-1884), where he served as resident engineer. He also supervised the construction of new dock works at Galway and at Sligo.

Price also carried out a variety of railway work, including acting as consulting engineer for the Cork & Macroom Railway. He was engaged by the Waterford & Limerick Railway and other companies in various parliamentary cases. He served for an academic year as deputy professor of civil engineering at TCD during the illness of professor Robert Crawford.

His 1879 paper to the ICE on 'Movable Bridges' won him a Telford Premium from the ICE and, in recognition of his work at Spencer Dock, he received a Master of Engineering (MAI) degree from his alma mater. He had previously received a Telford Medal & Premium for a paper in 1871 on 'The Testing of Rails'.

Price was a strong advocate of standard gauge light railways (rather than narrow gauge). He was retained by the Government as a commissioner to hold inquiries and report on various light railway and tramway schemes in the north and northwest of Ireland.

On 7 February 1855, Price married Frances Alicia Peebles, and had nine sons, including the engineer Alfred Dickinson Price, and two daughters.

Price was elected MICEI in 1861, served on the council for a number of years, was a vice-president, and served as president in 1895. He was elected MICE in 1870.

Price died on 4 April 1895 at his residence, "Knockeevin", Greystones, county Wicklow, of influenza and pneumonia, and is buried in Redford Cemetery in Greystones.

DILLON, James (1833-1916), civil engineering consultant, was born in Dublin in 1833, the son of Charles Dillon, a solicitor in practice with his brother John at 30 Arran Quay, Dublin. James was educated privately and then served an apprenticeship with a civil engineer Maurice Collis, MRIA from 1849 to 1853.

From 1853 to 1859 Dillon was resident engineer for the contractors for fifty miles of railway extensions in Cavan, Monaghan and at Fermoy. In 1859, he commenced private practice and was consulting engineer for the proposed Dundalk & Carrickmacross Junction Railway. (1861) and (with George Willoughby Hemans) for the proposed Dublin & Baltinglass Junction Railway. (1862). In 1866, he succeeded Sir John Macneill as Consulting Engineer to the Dublin & Meath Railway. He was also involved as the engineer on the construction of the Great Southern & Western Railway extension from Mitchelstown to Fermoy in county Cork.

During the 1870s, he designed and carried out many large river and arterial drainage works and flood regulating weirs in county Meath, and was responsible for the erection of a number of substantial bridges. The drainage areas included the Upper Inny, Stoneyford, Garristown and Baltracy rivers and two subdivisions of the proposed great Barrow River drainage, viz. the Kildare and Rathangan rivers.

In the 1880s, Dillon was appointed by the government to report on the engineering merits of several light railway schemes in Ireland and was also appointed government arbitrator in connection with the compulsory purchase of land and preparation of 'awards' required for railways and other large works. He was awarded a gold medal by the International Inventions Exhibitions 1885 for his patent hydrographic surveying and sounding apparatus.

Dillon served in 1905 on the Arterial Drainage Commission (Ireland) and was appointed by parliament to receive evidence and report on the subject of arterial drainage in Ireland, its then condition, and how best to maintain it.

He was elected MICEI in 1863, served as a member of council, and was president 1896-98. He was elected MICE in 1880.

In about 1873, Dillon married Esther Kenny and had a daughter Nora.

He died at his residence "Stratford House", Silchester Road, Glenageary, county Dublin on 12 May 1916, of a chronic chest infection, and is buried in Deansgrange cemetery.

WILSON, Wesley William (1840-1902), civil engineer, was born in county Louth on 2 March 1840, the son of William Charles Doyne Sillay Wilson, a retired HM Commissioner.

Wilson spent his entire career with the brewing firm of Arthur Guinness, Son and Co. of Dublin. Based at the St James's Gate headquarters, he joined the company in 1861 and trained from 1863 to 1867 under the company engineer James Bolger. Wilson was then appointed assistant engineer and in 1874 promoted to chief construction engineer with responsibility for all new work as well as the maintenance of the existing works (in essence Head of the Works Department). He went on to design warehouses, workshops, river quays, and a light railway system connecting the different parts of the brewery, including a spiral tunnel and heavy concrete works. He also designed improvements to Lord Iveagh's estate nearby at Palmerstown, including a 180ft span wrought-iron lattice girder bridge spanning the river Liffey. Wilson retired from Guinness's, aged 60, on 8 April 1899.

On 1 January 1872, at St George's, Dublin, Wilson married Kathleen Margaret, daughter of Hazlet Hamilton. They lived at a number of addresses within easy reach of the brewery, but by the time of the 1901 census, he is recorded as residing at 120 Merrion Road in Ballsbridge, with his Donegal-born wife and two daughters, one an accountant, the other attending college. A son, Charles William, was born 3 December 1874.

Wilson was elected AMICEI in 1875 and transferred to MICEI in 1882. He was a member of council from 1887, a vice-president 1895-97 and served as president 1898-99. He assisted with the preparation of the contract drawings and the carrying out of the building of the headquarters of the Institution at 35 Dawson Street and made a significant contribution to defraying the expenses incurred in the undertaking. He was elected AMICE in 1893. He was also a member of the RIA, having been elected in 1888.

He died suddenly and unexpectedly of chronic hepatitis at his residence "Ardganagh", Ballbridge, Dublin on 3 May 1902.

GLOVER, Edward (1847-1931), civil engineer, was born at Maghowan, Churchtown, Buttevant, county Cork on 9 December 1847, the son of James Glover of Maghowan and his wife Mary from county Limerick.

Glover was educated at Bandon Grammar School and at Trinity College Dublin, where he entered in January 1870 to study civil engineering. He received his BAI degree in 1874 from the University of Dublin and an MA in 1877. Following graduation, he served a pupilage with Peter Burtchaell, the county surveyor of Kilkenny, and travelled to France, Germany and Belgium, before being appointed county surveyor to the southern district of county Mayo in November 1876, where he had control of all public works, including nearly 1,000 miles of roads. He designed the original road connection from the mainland to Achill Island, consisting of two causeways joined by a 120ft-span light girder swing bridge (pivoted on the central pier) over the neck of Achill Sound. The bridge was completed, with some modifications, by his successor Peter Cowan and was opened in 1887. Glover was engineer and architect to the board of Mayo lunatic asylum from 1877 to 1893 and planned a substantial enlargement to the facilities at Castlebar. He designed and carried out sewerage works for Ballinrobe, Castlebar and Westport, and a waterworks scheme for Westport. Glover also reconstructed the pier at Clare Island and the quay walls at Newport.

Moving to Kildare in January 1886, Glover took on responsibility for the maintenance of some 1,150 miles of public roads, and built a number of bridges. With Henry Vincent White, the county surveyor of Offaly, Glover was heavily involved with the organisers of the 1903 Gordon Bennett Cup motorcar races in the selection and preparation of the route. The surveyors agreed to have the relevant roads steamrolled, widened and generally improved, the international event being deemed an unqualified success due largely to their efforts. Glover became ill towards the end of 1913 and took retirement at the end of June the following year.

Glover married firstly Sarah Mary, daughter of Henry Brewster, surveyor for North Mayo, by whom he had one daughter, and secondly, in about 1884, Margaret, the daughter of Luke Knight from Clones in county Monaghan, and widow of Walter Young from Gartinadress, county Cavan. Glover lived with his family at Prince Patrick terrace, North Circular Road, Dublin for most of his term of office in Kildare, and until his death in 1931.

In June 1893, he was elected a fellow of the Royal Institute of British Architects, and was in private practice with his brother-in-law, Charles Astley Owen, FRIBA, from 1887 onwards with offices in Molesworth Street in Dublin.

He was elected MICEI in 1877, served on council 1894-97 and from 1902 until his death, was vice-president 1898-99, and president 1900-02. He was also elected AMICE in 1885 and transferred to MICE in 1891.

Glover died at Portobello House nursing home in Dublin on 20 March 1931.

RYAN, John Henry (1846-1929), civil engineer, was born at Clonmel, county Tipperary on 5 January 1846, the younger son of Thomas Ryan of Killeffernan House, Clonmel, county Tipperary, by his first wife, Mary Grace, daughter of John Hewetson of Castlecomer, county Kilkenny. He was educated privately and entered Trinity College Dublin in November 1864, where he studied civil engineering. He received the degree of BA from the University of Dublin in 1868 and, in the same year, a Licence in Civil Engineering (LCE). He was awarded an LL.B degree from the university in 1890.

He served a short pupilage with Alexander McDonnell at the Great Southern & Western Railway locomotive works at Inchicore and with George Leeson, before moving to the USA. Here he was in private practice as a civil engineer and also held the position of divisional and resident engineer on the Burlington & Missouri Railroad in Nebraska for a period of one year, on the Chicago & North Western Railroad for two years and on the Texas & Pacific Railroad for six years.

Following his return to Ireland in 1880, Ryan worked for James Dillon on the river Inny, Rathangan, Garristown and Ward river drainage works. He also acted from time to time as an inspector for the Board of Works (Ireland) under the tramways and other acts. He then returned to private practice at 22 Nassau Street in Dublin and also maintained an office in Victoria Street in London.

In 1889, Ryan was appointed joint engineer with Edward Townsend for the 48-mile single-track branch line from Galway to Clifden for the Midland Great Western Railway. The line included the Corrib Viaduct at Galway and was opened in 1895 (closed April 1935). Ryan was also engineer to the Tralee & Dingle Railway (reopened to Blennerville as a heritage line, but currently not operating), the Kenmare & Headford Railway, the West Kerry Railway (Killorglin to Valentia Harbour), and a proposed light railway from Castlecomer to Athy (not proceeded with).

In 1902, he was appointed chief engineer to the Hudson's Bay & Pacific Railway (from Churchill to Calgary), although he appears to have acted only in a consulting capacity from his office in Dublin. At the time, he was living in Waterloo Road in the city.

Ryan was much in demand as a government arbitrator for the Local Government Board for Ireland, the Board of Works (Ireland) and the Board of Trade. He was a member of the Viceregal Arterial Drainage Commission and, as acting inspector for the Board of Works, held inquiries into a number of drainage schemes. He was a member in 1907 of the Dublin International Exhibition committee and was co-author, with George Coppinger Ashlin, of the initial report on the project.

On 8 November 1887, Ryan married Henrietta Anne, daughter of William Stewart and Jane Bellingham, Ravensdale, County Kildare and The Cliffs, Howth, county Dublin.

Ryan was elected MICEI in 1879, re-elected in 1896, was a member of council 1898-1928, was honorary treasurer 1906-10, vice-president 1899-1902, and served as president 1 May 1902-03. He was elected MICE in 1884, and represented Ireland on council in 1907.

Ryan closed his Dublin practice in 1920 and retired to the family home, Killeffernan House in Clonmel, where he died, of heart disease, on 10 May 1929, predeceased by his wife.

COCHRANE, Robert (1844-1916), architect, was born at Inch, near Downpatrick, county Down on 21 July 1844, the son of Hugh William Cochrane and Mary Ann Williamson. Robert was educated at Royal Belfast Academical Institution and entered Queen's College, Belfast in October 1864. After the first year of his BA degree course, he entered the engineering school and in 1868 obtained a Diploma in Engineering.

On graduating he became articled to Henry Smyth, county surveyor of Down, and served as town surveyor of Banbridge and Dromore and assistant county surveyor for the Banbridge district. He also acted as engineer and surveyor for the county Down estates of the Earl of Clanwilliam and for the trustees of General Meade.

In 1874, having come first in a competitive examination, he was appointed to the architects' department of the Board of Works (Ireland) and moved from Banbridge to Athlone as clerk of works (later known as assistant surveyor of buildings) for the western district of the country comprising counties Galway, Mayo, Roscommon and Sligo. He lived at Excise Street in Athlone. In 1887 he was promoted to principal surveyor in the Board's Dublin headquarters at 51 St Stephen's Green. During his time at the Board of Works, he specialised in the design of many post-office buildings and coastguard stations and designed extensions to Queen's College Belfast, including the Hamilton Tower and the Students' Union buildings. Cochrane survived a severe illness in 1907 before retiring from the Board of Works in 1909.

In 1879, Robert married Ethel Mary Sarah, daughter of George Hawken of Battersea and had two sons and a daughter. One of his sons, Hugh Carew Cochrane became a civil engineer.

Robert Cochrane was elected AMICEI in 1870 and transferred to MICEI in 1889. He served on the council 1898-1900, was vice-president 1901-1903, and served as president 1904-1906. He also served for a period as vice-president of the Engineering and Scientific Association of Ireland. In 1905 he received an honorary LL.D. degree from the Royal University of Ireland. He was elected a member of the Royal Irish Academy in 1885 and a fellow of the Royal Institute of British Architects in 1892.

In addition to his professional interests, Cochrane was a highly regarded and meticulous antiquarian scholar. As a fellow of the Royal Society of Antiquaries of Ireland (RSAI) from 1864, he was largely responsible for the regeneration and reorganization of that society, of which he was president from 1909 until 1912. A combination of his antiquarian and architectural expertise made him ideally suited for the post of Inspector of Ancient and National Monuments for Ireland, to which he was appointed in December 1899 and which he held until his death.

Robert Cochrane died of a brain haemorrhage at his residence, 17 Highfield Road, Rathgar, Dublin on 17 March, 1916.

ROSS, William (1855-1926), civil and mechanical engineer, was born in Dublin on 14 February 1855, the youngest son of William Ross, engineer and founder of the firm of Ross & Murray in the city. He was educated at Blundell's School, Tiverton and, like his older brother George, at Merchiston Castle School, Edinburgh before starting his engineering career by serving a four-year apprenticeship at the North Wall iron works of Ross, Stephens & Walpole in Dublin. He entered Trinity College Dublin in April 1872 to study civil engineering. He graduated with the degrees of BA and BAI in 1876 from the University of Dublin and received an MA degree in 1879.

Following graduation, he became a partner in Ross & Walpole and was chairman and managing director until his retirement in 1923. Work undertaken by the firm included iron bridges for the Midland Great Western Railway and the Dublin, Wicklow & Wexford Railway. He was chairman of Dalkey (county Dublin) UDC in 1912, and chairman of the Dublin Port and Docks Board. He resided at 25 (and later at 9) Dalkey Avenue in Dalkey.

In about 1883, he married Scottish-born Mary Jane Turner.

Ross was elected MICEI 1884, was a member of council 1891, 1894, 1907-1926, honorary treasurer 1895-1905, vice-president 1902-05, and president 1906-07. He was elected AMICE in 1894. He was also a member of the Institute of Naval Architects and a member of the IMechE, elected 1881.

William Ross died at "Summerfield", Dalkey, county Dublin on 26 October 1926 and is buried in Deansgrange cemetery.

He died suddenly on 26 October 1927 at "Summerfield", Dalkey, county Dublin, the former residence of his elder brother William, who had died the previous year.

MOORE, Joseph Henry (1844-1912), civil engineer, was born in Westminster, London on 26 August 1844, the eldest son of Rev.Patrick Moore, chaplain at the Training College, Westminster, and Ellen Mary Ashe. Joseph was educated at Drogheda Grammar School and at Lewisham Grammar School in Kent before entering Trinity College Dublin in 1862. Here he was awarded a scholarship in 1865 and a gold medal, graduating BA in 1866 and receiving a Licence in Civil Engineering in 1867. He proceeded to the degree of MAI in 1874.

After completing his university studies, Moore served a pupilage with Bindon Blood Stoney at Dublin port and later worked with Beyer, Peacock & Co. in London.

Having worked as assistant county surveyor in Antrim under Alexander Tate, he took first place in a county surveyor examination in November

1870 and was appointed to Westmeath in the following month. He transferred to county Meath in August 1874 at his own request and served in Meath until his retirement in 1907.

Moore was an early user of a form of reinforced concrete for bridge works. In the early 1880s, he built bridges of small span over the river Nanny using disused rails with nine inches of concrete laid on top. In 1903-4 he was responsible for the design and construction of a four-span bridge over the river Boyne at Watergate, Trim. The spans consisted of a grillage of steel beams with concrete between. The bridge was replaced in 2004 with a single-span low-rise steel arch structure.

On 14 June 1871, Moore married Elizabeth Jane, daughter of Rev.Robert William King, Church of Ireland rector of Portglenone, county Antrim, with whom he had four sons and four daughters.

He was elected AMICEI in 1863, MICEI in 1874, was a member of council 1891, 1898-1906, vice-president 1906-07, and served as president in 1907. He then became honorary treasurer of the ICEI until his death.

He resided at 123 Anglesea Road in Dublin, but died of cancer at the Meath Hospital on 18 March 1912 and is buried at Mount Jerome cemetery in the city.

ROSS, George Murray (1852 - 1949), civil engineer, was born in Dublin in 1852, the eldest son of William Ross, the founder of the firm of Ross & Murray, engineers. George was educated at Merchiston Castle School in Edinburgh and entered Trinity College Dublin in 1869 to study civil engineering. He graduated with the degrees of BA and BAI in 1873 from the University of Dublin and received an MA degree in 1877.

After graduating, he first joined the staff of the Board of Works as an assistant surveyor, but shortly afterwards joined his father's business. He later closed down the business and set up as an independent consultant at 61 Dawson Street in Dublin. In 1898-9 he entered into partnership with William Kaye-Parry, specialising in domestic sanitation. He is recorded at a number of business addresses in the city and lived variously at Ballinagowan House in Rathmines, in Donnybrook and at Dalkey.

On 8 October 1879, George, by then a widower, married Alice Jane, daughter of William Johnson of Prumplestown House, Castledermot, county Kildare. A son George Mabyn Ross, born 1884, had a distinguished career as an engineer in India.

George Murray Ross was elected an associate of the ICEI in 1874, served on the council 1900-1907 and 1911-1927, was a vice-president 1907-1909, and served as president 1909-1911. As president he acted as vice-chairman of the First Irish Roads Congress (1910) and delivered a paper on 'water supplies for small areas' at the Congress of the Royal Institute of Public Health the following year. During WW1 in 1917 he went to France as senior engineer in charge of a labour battalion for the construction of roads and railways. He was elected a member of the RIAI in 1925.

COWAN, Peter Chalmers (1859-1930), civil engineer, was born at Peters Street, Murraygate, Dundee on 20 March 1859, the son of James Cowan, a merchant, and Mary Ann Chalmers. Peter was educated at The High School, Dundee and was then articled to a civil engineer, Robert Blackadder. In 1878, Cowan entered the University of Edinburgh and graduated in May 1881 with a BSc degree. In 1909 he received an honorary degree of DSc from his alma mater.

For a year after qualifying, Cowan acted as an assistant to Professor Jenkin at the university and, having being awarded a Vans Dunlop Scholarship, a three-year scholarship allowing him to study at the university of his choice, travelled to the USA. It is not clear whether he continued his studies as in 1882 it is recorded that he served for some months under the New York City Surveyor, T.G.Smith. He then became an assistant engineer on the New York, West Shore & Buffalo Railway (obtaining a Miller Prize from the ICE for a paper on his work), and later with the Canadian Pacific Railway responsible for track construction through difficult terrain at the north-west corner of Lake Superior. Following his return to England in November 1884, Cowan became assistant engineer on railway and dock works with Alexander Cunningham Boothby, MICE and John Macrae, MICE.

Moving to Ireland, Cowan was appointed, after a competitive examination, county surveyor for South Mayo in February 1886. This entailed the maintenance of roads and bridges, piers and courthouses, and some involvement in lines of new railways. He was also employed by the sanitary authorities in regard to water supply and sewerage. He acted as consulting engineer to the Piers and Roads Commission (1886-7). With James Perry, county surveyor for the western district of County Galway, Cowan assisted in the design and supervision of 86 separate projects, mainly involving small harbours and piers, as well as some new roads and bridges. During this time, he lived at The Mall in Westport.

In June 1889, Cowan was appointed county surveyor for South Down and in March 1890 was given responsibility for the whole of the county. His work in Down was notable for his efforts, along with a number of assistants, to improve the state of the 2,800 miles of road in the county. He claimed in 1891 that the system of management of county works in Ireland was twenty years ahead of the systems in England and Scotland, with all the works, and every road, under the control and supervision of a professional officer, appointed after a competitive examination. Cowan was a strong advocate of, and the first in Ireland, to introduce steamrolling of roads.

On 23 January 1899, he was appointed Chief Engineering Inspector to the Local Government Board of Ireland (LGB), in succession to Charles Philip Cotton, and held the post for twenty years. Following the enactment of the Local Government (Ireland) Act 1898, there was a great expansion in the activities of the Board, a rapid development in road traffic, the introduction of direct labour and new road making techniques, an expansion of public housing programmes, and accelerating activity in the areas of public water supply and waste disposal.

Cowan held numerous public inquiries into local authority proposals under all the above headings and also contributed a number of papers on aspects of road making to conferences and in the published literature. He was an official general reporter at the International Road Congress held in London in June 1913.

In January 1918, he made a valuable report to the LGB on housing conditions in Dublin and recommended major housing projects for Dublin along the lines of a 'garden city'. From 1919, he was technical adviser to the Housing Committee Board. Following a directive from the Minister for Local Government in the provisional government to reorganise the department, Cowan's appointment was terminated abruptly in 1923 and, having negotiated a pension of £720, he went to live in England.

On 30 August 1888 Cowan married Marion, the daughter of Alexander Johnston, a Westport, County Mayo doctor. They had a daughter Hilda and three sons, two of whom were killed in action during WW1 whilst serving with the Royal Flying Corps, the other, Frederick Alexander Cowan, also becoming a civil engineer.

When in Belfast, Cowan lived at College Gardens in the city and, in Dublin, first at 34 Ailesbury Road in the south of the city, and later at "Castlemount", Castleknock in county Dublin.

Cowan joined the ICE as a student member in 1880, was elected AMICE in 1885 and transferred to MICE in 1892. He was a member of council 1921-1924. Elected MICEI in 1900, he served on the council, was for a time a vice-president, and served as president 1911-1913. He was also president of the Engineering and Scientific Association of Ireland 1915-18 and was an honorary fellow of the Royal Institute of the Architects of Ireland.

Cowan died at "Glendaragh Cottage", Church Road, Fleet, Hampshire on 9 August 1930 of a brain haemorrhage.

COLLEN, William Garibaldi (1861-1932), civil engineer, was born in Tasmania on 24 August 1861, the son of John Collen and Mary Harvey. John Collen was founder and principal of Collen Bros. Ltd. of Dublin and Portadown, which carried out many large building and construction contracts throughout Ireland, including the Royal Dublin Society's main hall (1884), the asylum at Portrane (1903), as well as numerous railway contracts and water supply schemes.

William was educated at Belfast Methodist College and entered Trinity College Dublin in June 1878. He obtained a BA degree in 1882, a BAI in 1884, and an MA in 1886, all awarded by the University of Dublin. He was a gold medallist at the university and distinguished himself in classics and mathematics in addition to his civil engineering studies.

For the first seven years after graduation, William gained experience of railway and other engineering work. He worked during 1883 on parliamentary surveys for James Price in county Meath, and for John Henry Ryan in county Kilkenny. He was employed 1884-6 under William Hemingway Mills as an assistant contractor's engineer on the Carrickmacross branch of the GNR(I) railway. During the period 1886 to 1889, he was employed by James Barton on sections of the Cavan, Leitrim and Roscommon Light Railway. Following this he worked for six months at Sligo Harbour.

In October 1891, Collen obtained fifth place in the County Surveyor's (Ireland) examination and bettered this in March 1892 by gaining first place. On 31 December 1891, he was appointed a district surveyor for the northern district of Dublin and from 1897, following a provision in the County Surveyors Act (1897), assumed sole responsibility for the whole of the engineering business of the county of Dublin. His offices were first at 9 Hume Street and later at 11 Rutland Square in Dublin. He resided at 78 Upper Leeson Street.

One of his most important contributions was the improvement of the roads in the county. The roads had become notoriously bad as levels of traffic increased with the introduction of the motorcar. He was strongly opposed to the contract system of road maintenance, advocating instead the use of direct labour. He also introduced steamrolling, even purchasing a steam roller with his own funds to convince the council of the effectiveness of the new system. By 1900, he had eleven rollers at work throughout the county and there was a rapid improvement in the state of the roads.

As county surveyor, Collen was also responsible for the maintenance of the small harbours at Rush and Lambay in the north of the county. Between 1905 and 1909, he planned and supervised the extension and reconstruction of Loughshinny pier. With John Smith as resident engineer, he was responsible in 1904-07 for Newtown and Lissenhall bridges at Swords, for coastal protection works at Barnageeragh near Skerries, and the Skerries water supply.

According to family records, William Collen married an Ida Brown, but there does not appear to be any official record of the event.

Collen was elected MICEI in 1902, served on the council 1908-1928, was vice-president 1909-13 and president 1913-15. He was elected AMICE in 1894 and transferred to MICE in 1898.

On taking early retirement in June 1924, William Collen moved to Bournemouth, where he did not enjoy good health. He died at his residence in Bethia Road, Bournemouth on 29 April 1932 following a severe illness.

RUDDLE, Mark (1854 - 1919), civil and electrical engineer, was born Marcus Ruddle at 6 Haddington Terrace, Dublin on 27 March 1854, the son of Marcus Ruddle, civil engineer and his wife Catherine. His father was a clerk in the Valuation Office and his mother was from Broseley in Shropshire.

He received his early education under the tutorship of George Porte, MRIA from whom he gained an insight into the science of engineering. He served a pupilage from 1870 to 1874 with John Douglas, mechanical engineer, in Portsmouth and from 1874 until 1878 acted as chief engineer at the Stanisham Chemical Works in a suburb of the city.

Between 1879 and 1881, Ruddle was an assistant engineer at the Edison Telephone Company and with the United Telephone Company of London. The following year he worked as an assistant electrical engineer with the Edison Electric Light Company in London, where the first system of underground cables was laid down under his direction. In 1882 he became Chief Electrical Engineer for the Edison group, based in Manchester. In 1889 he served for a year as a manufacturing technician and then became assistant engineer with the Metropolitan Electric Light Company in Manchester and in London. In 1881 he is recorded as living with his widowed mother at 47 Beulah Road in Walthamstow, Essex.

In 1891, Ruddle moved back to Ireland to represent Dublin Corporation interests during the erection of the first electricity generating station at Fleet Street in the city, the consultant being E.Manville and the contractors the Electrical Engineering Company of Ireland in association with Messrs Hammond and Company of London. On completion of the contract, Ruddle was appointed electric lighting superintendent to Dublin Corporation. He was later to be appointed to the new post of Dublin City Electrical Engineer and lived at 45 London Bridge Road in the city.

The load on the Fleet Street station increased rapidly and it was decided to erect a new power station at Pigeon House Harbour as part of a completely new electrical supply system for Dublin. Ruddle worked with consulting engineer Robert Hammond on the coal-fired power station, which when commissioned in 1904, had a capacity of 3MW. The electricity was generated by three-phase alternators at a pressure of 5,000 volts at 50 cycles, and transmitted at that voltage by underground trunk cables to a distribution station on the site of the old Fleet Street station, whence it fed a three-phase supply network throughout the city. This represented a world first in the supply of a three-phase electricity supply to an urban area.

In 1906 Ruddle was joined by Laurence Kettle, who was appointed as an electrical and mechanical engineer to the Pigeon House station, the capacity of which had increased to 12MW by the time of Ruddle's retirement, due to ill-health, in 1919.

On 18 June 1902, Ruddle married Adelaide Mary Louisa, daughter of the late Samuel Irwin, in Rutland Square church in Dublin.

Ruddle was elected MICEI in 1896, served on council, was a vice-president and was elected president in 1915 (the first electrical engineer to serve in that position). He was also a member of the IEE, having been elected an associate in 1892, AMIEE in 1899, and MIEE in 1902. He served for a number of years on the committee of the IEE Dublin Section, becoming chairman in 1904-5.

He died of pneumonia at his residence, 7 Landsdowne Terrace (Shelbourne Road), Dublin, on 3 August 1919, survived by his wife and son.

LILLY, Walter Elsworthy (1867-1940), mechanical engineer, was born at 51 Devonshire Road, Islington, London, the son of William Elsworthy Lilly, a drapery warehouseman, and Lydia Margaret Leslie. He was educated at Aske's Hatcham School, Brockley, and at Lewisham High School, before being articled from 1882 to Stephens & Woodman at their Brixton engineering works, and later in March 1884 to George Walter & Co. of Blackfriars. In 1885 he moved to the Whitechapel engine works of Bringes & Goodman, and two year later, was appointed assistant manager of the Albert Dock engine works.

In 1891, Lilly was appointed assistant to Thomas Alexander, professor of civil engineering at Trinity College Dublin. Lilly was responsible for developing the mechanical engineering aspects of the general engineering course and initiating a programme of scientific engineering research. He had become a Whitworth Exhibitioner in 1891, but was conscious of his lack of a primary academic qualification. So, in 1898, Lilly was allowed to take the final examinations for the BAI degree, which he passed with flying colours. Three years later he received the degrees of MA and MAI, both from the University of Dublin. He received the degree of ScD in 1907 from his alma mater, the year that he was appointed to a new lectureship in mechanical engineering and given the task of designing and laying out the mechanical engineering workshops in the School of Engineering. During WW1 he was Civilian Inspector at Woolwich (1916-1918), for which work he was awarded an MBE.

Lilly was a clever investigator and wrote on such diverse subjects as the design of arches, the elastic limits and strengths of materials, the gyroscope, the collapsing pressure of circular tubes, the design of hollow shafts, pump valves, marine boilers, and the design of plate girders. But it is in connection with his work on the strength of columns that Lilly is best known. Although heavily handicapped by lack of equipment, having at his disposal only a ten-ton testing machine, he conducted a long and arduous series of experiments and expressed his conclusions in a series of papers for which he was awarded gold and silver medals by the ICEI and a Telford premium by the ICE. During his time at Trinity, where he lived in rooms in No.39 in New Square, he published over thirty scientific and professional papers. His main published treatises dealt respectively with the design of plate girders (1904), the design of columns, girders, and shafts (1908), and the design of arches (1915).

Lilly was elected MICEI in 1901, acted as honorary secretary 1906-09, served on council 1909-10 and 1912-13, vice-president 1913-17 and was president 1917-18. He received honorary membership in 1923, the year that he retired and went to live near Cromer in Norfolk. He was also an associate member of the IMechE, elected 1902.

Dr Lilly died, unmarried, at East Runton, Norfolk on 3 January 1940.

MOYNAN, John Ouseley Bonsall (1854-1932), civil engineer, was born in Dublin about 1854, the eldest of eight children of Richard Moynan, a manager with Ferrier Pollack & Co., and Harriet Noble, daughter of a Church of Ireland clergyman.

John was educated privately at Nice before entering Trinity College Dublin in June 1869 to study civil engineering. He received the degrees of BA in 1873 and BAI in 1874 from the University of Dublin and an MA in 1890. He served a pupilage with James Price and was resident engineer in charge (1876-78) of the works on the first division of the Waterford, Dungarvan & Lismore Railway and was later employed on various projects under county surveyors John Henry Brett (Kildare) and Edward Glover (Mayo).

In September 1883, Moynan was appointed county surveyor of Longford, but he was not happy with his position there and, in August 1891, transferred to Tipperary (North Riding), where he was to serve for the next thirty-nine years.

He was something of a pioneer in the introduction of large-scale direct labour road schemes and was amongst the first to advocate and use steamrolling extensively. Beginning in 1903 with a scheme under which up to 350 miles of the more important roads were dealt with by direct labour, all work on the county's 1,300 miles of roads had been transferred to the new system by 1909.

In the overall national context, Moynan saw little future in the canals and very limited prospects for extensions of the railway network and he correctly forecast that roads would become the dominant mode of transport, especially for freight. He paraphrased a well-known couplet to read: *"Canals were, railways are, roads will be, the most important of the three"*. Moynan also recommended transferring responsibility for some 25% of rural roads linking the more important towns to a central authority and classifying them as main roads and bringing them up to a uniform standard over a period of years.

Between 1910 and 1915 he supervised the construction of a replacement road bridge across the river Shannon at Portumna. This 30ft-wide steel plate-girder bridge of six spans of 80-90ft resting on 9ft diameter cast-iron cylindrical piles was designed to Moynan's specification by Charles Edward Stanier of London and completed in 1911.

In Nenagh, Moynan developed an extensive private practice in association with the assistant county surveyor, Robert Paul Gill, operating from 1903 under the title Moynan & Gill, Architects. Work included Nenagh town hall and the extension of the town's water supply He did not believe that there was much future in developing the water resources of Ireland and argued that the proposed Shannon Scheme would not be financially or economically possible. In December 1929, just after the Shannon Scheme had come on stream, Moynan resigned as county surveyor, but continued in post until the following September.

Moynan's second marriage, on 22 July 1886, was to Henrietta Elizabeth Catherine, daughter of John Knox of Dublin.

Moynan was elected MICEI in 1881, served on council 1909-17, 1918-1932, was vice-president 1917-18, and served as president 1918-19. In 1886 he was awarded a Mullins Silver Medal for a paper on 'The maintenance and repair of county roads in Ireland'.

Moynan died on 9 October 1932 at Islandbawn House, Nenagh, county Tipperary, of a brain haemorrhage.

McCARTHY, Patrick Hartnett (1868-1942), civil engineer, was born at Glenmore, near Newcastle West, county Limerick on 1 December 1868, the son of Patrick McCarthy, a farmer, and Bridget Hartnett. He was educated privately and later at university. He entered the office of Samuel Gordon Fraser in February 1889 and spent the next ten years with Fraser, first as a pupil, and then as an assistant, during which time he was engaged in the planning, promotion and construction of a number of railways, light railways and tramways, including the Ballina & Killala Railway, the Collooney & Claremorris Railway, Hill of Howth Tramway and the Dublin Southern Tramways. He was resident engineer on the construction of the Dublin to Dalkey and other tramway lines.

Whilst still working for Fraser, McCarthy, in October 1897, at the age of 28, entered Trinity College Dublin to study civil engineering. He obtained the degrees of BA and BAI from the University of Dublin in June 1901 and received an MA degree in December 1904.

With his long experience in taking-off earthwork quantities for railway construction, McCarthy recognised the need for reducing the labour involved in such work, and devised and made an ingenious instrument, somewhat similar to a planimeter, which enabled the volume of cuttings to be taken off from longitudinal sections.

In January 1899, whilst still studying at college, he established a consulting engineering practice in Dublin that became in time a countrywide practice, joined in 1925 by his son, Patrick Joseph McCarthy. The main activity of the practice was in connection with the planning and construction of waterworks and sewerage schemes. Amongst the schemes completed were the water supplies to Arklow, Gorey and Courtown in county Wexford, Monaghan, Laytown-Bettystown in north county Dublin, and Mallow, and sewerage schemes for Foxrock, Dundrum, Stillorgan and Terenure in south county Dublin, Howth, Roscommon, Mullingar, and Skerries. From 1912 he acted as Engineer to the Rathdown No.1 Rural District Council (essentially south county Dublin).

McCarthy's extensive knowledge of parliamentary procedures led to him being retained in connection with the preparation of plans and estimates for the London United Tubular Railways. For many years he was consulting engineer to the standard Irish gauge Dublin & Blessington Steam Tramway (opened 1888, extended to Poulaphuca 1895) and from 1910 consulting engineer and architect to Bray UDC. He was retained as consultant by the Dublin Port & Docks Board to report on the improvement and extension of the port to accommodate increased overseas traffic. He was also responsible for carrying out a harbour improvement scheme at Dundalk. He was retained in an advisory capacity on the preparation of the contract documents for the Shannon Power Development (the Shannon Scheme) and afterwards on the many technical problems arising during its construction (1925-28).

McCarthy acted as external examiner in civil engineering for the National University of Ireland (1913-15) and was a member of the Canal & Inland Waterways Commission (1922-23) that examined and reported on the canal system in Ireland for the Provisional Government.

On 27 June 1900, McCarthy married Mary Josephine, daughter of Roger Greene, Solicitor, of Dublin.

McCarthy was elected MICEI in 1904, became a member of council in 1908, was vice-president 1917-19, and served as president 1919-20, afterwards remaining on the council as a past-president until his death.

McCarthy lived at "Beechfield", Bray, Co.Wicklow, but died suddenly at his practice at 26 Lower Leeson Street, Dublin on 29 April 1942.

BERGIN, Francis (1855-1925), civil engineer, was born at Kildare on 21 November 1855, the son of Charles Bergin, a farmer, and Mary Lee. Francis was educated at Tullabeg College, Tullamore and entered Trinity College Dublin in November 1873 to study civil engineering. He graduated with a BA degree from the University of Dublin in 1878 and a BAI degree in 1879.

In December 1880 he emigrated to New South Wales (NSW), accompanied by his first wife, Nancy, and their two-month-old son Charles. Two daughters, Alice and May, were born in Australia. He spent the next seven years as an engineer with the NSW government railways.

On his return to Ireland in 1887, he took time to pass the Civil Service examinations qualifying him for appointment to a county surveyor post. However, he did not seek an appointment as a county surveyor. Instead, from 1887 to 1890, he acted as resident engineer on the Barrow drainage survey and was subsequently employed until 1893 as engineer to the Grand Canal Company.

From 1894 until his death he was engineer to the Naas (County Kildare) Union RDC, and in 1900 was elected assistant surveyor for county Kildare. He was also Town Surveyor for Kildare in 1901. He lived at 1 The Square and later at 7 Dublin Street in Kildare.

He practised as an engineering consultant from offices at 36 Westmoreland Street in Dublin and designed water and sewerage schemes for many towns in Ireland. He also acted in 1906 as consulting engineer to Limerick Corporation for the extension of the city's waterworks and for Dublin Corporation for the large supplementary reservoir near Roundwood in county Wicklow.

His first wife, by whom he had two more daughters after the family returned to Ireland, died while her children were still young. On 20 September 1905 Bergin married his daughters' governess, Mary (Minnie), daughter of Patrick Murphy, a merchant and farmer from Wexford, the union resulting in the birth of a further two sons and two daughters.

Bergin was elected MICEI in 1892, was a member of council 1907-1909, 1910-1911, 1913-1923, vice-president 1918-19, and served as president 1920-21. He was also a member of the Engineering Association of New South Wales (est. 1870), one of a number of associations that came together in 1919 to form the Institution of Engineers, Australia.

Francis Bergin died at his residence, "Beechgrove", Kildare on 26 October 1925 of cancer of the liver.

HARGRAVE, Joshua Harrison (1860-1924), civil engineer, was born at Wrixon's Place, Cork on 15 July 1860, the eldest son of Abraham Addison and Isabella Hargrave. In 1876 Joshua entered Queen's College Cork to study civil engineering, graduating from the Royal University of Ireland in 1879 with a BA degree and in October 1881 with a BE degree with honours. From 1882 he served a two-year pupilage with Wells-Owen & Elwes of Westminster, London.

From December 1883 he was on the engineering staff of the Great Western Railway at Paddington under the Chief Engineer, William George Owen (1810-1885), before returning to Ireland in 1889 to take up an appointment as draughtsman with the Great Northern Railway of Ireland (GNR(I)). He later became chief assistant to William Hemingway Mills, Engineer-in-Chief of the GNR(I), succeeding him as Chief Engineer in about 1903. Hargrave remained with the company until his death in 1924, being occupied chiefly with bridge construction and the laying out of railway extensions, including the Hill of Howth electric tramway, which was operated by the GNR(I) from 1900.

On 12 June 1897, Hargrave married Louisa, second daughter of John Foster Newland, a medical doctor, of "Mount Haigh", Kingstown (now Dun Laoghaire).

Hargrave was elected AMICE in 1886 and transferred to MICE in 1904. He was elected MICEI in 1898; served on the council 1904-1920; was honorary secretary and honorary treasurer at different times between 1913 and 1923; vice president 1920-21; and served as president 1921-22.

Hargrave was a keen yachtsman, being at various times vice-commodore, honorary secretary and honorary treasurer of the Dublin Bay Sailing Club. He served the Water Wags Club for many years and was untiring in his work for the Dun Laoghaire branch of the Royal Lifeboat Institution, including serving as its secretary.

He died at his residence 4 Haddington Terrace, Kingstown (Dun Laoghaire) on 24 January 1924 of stomach cancer.

PURCELL, Pierce Francis (1881-1968), civil engineer, was born at Kilkenny on 6 October 1881, the son of Thomas Patrick Purcell, Chief Officer, Dublin Fire Brigade, and Margaret Phelan. He was educated at St Vincent's College, Castleknock (1897-99) and at Trinity College Dublin (1899-1902), where he was a Senior Moderator and Gold Medallist and studied civil engineering. In 1903 he was conferred with the degrees of BA and BAI with honours by the University of Dublin, the degrees of MA and MAI following in 1908. He received an honorary DLitt from the National University of Ireland in 1952.

After a brief spell of railway survey work in county Meath, notably the Mullingar, Kells & Drogheda Railway, Purcell joined the works department of the London County Council in 1904 as assistant engineer under Sir George William Humphries and was engaged on the Holborn-Strand improvements, including the construction of the greater part of Kingsway and Aldwych. When this work was completed in 1906, he became assistant resident engineer under Sir Maurice Fitzmaurice on the construction of the south London main drainage tunnels.

In 1909, Purcell was appointed first professor of civil engineering at the newly established University College Dublin and faculty dean. He was a member of the governing body of the college and, as chairman of the buildings committee, played a major part in the acquisition of properties as a site for the present campus at Belfield in Dublin. He carried out research into cement and concrete mixtures and took a leading role in the drafting of the first Irish standard for Portland cement.

He took an active part in the practical development of the natural resources of Ireland. In 1917, he became secretary to the Irish Peat Enquiry Committee, set up by the Department of Scientific and Industrial Research (DSIR) under the chairmanship of Sir John Purser Griffith and in 1920 was appointed peat investigation officer to undertake research into the mechanical harvesting of peat and its use as a boiler fuel. He was director of Ireland's first peat briquette factory at Lullymore in County Kildare. Professor Purcell was author of *The Peat Resources of Ireland*, published by DSIR in 1920.

In a later role as a consultant, he was responsible for two substantial reinforced concrete bridges, one at Kenmare, County Kerry and the other Butt Bridge in Dublin. He was also involved as consultant for a wharf at Waterford, regional water supplies in Counties Carlow and Wicklow, as well as a number of sewage schemes.

On 30 June 1910, at St Joseph's, Highgate Hill, London, Purcell (Roman Catholic), married Australian born Amy Austral (Church of England), daughter of George Henry Oatway, a company director and electrical engineer. Their son Pierce Michael Oatway Purcell was also a civil engineer.

Purcell was elected MICEI in 1911, was honorary secretary 1912-15, council member 1915-1920, 1924-1961, vice-president 1921-22, and president 1922-24. He was elected AMICE in 1905 and transferred to MICE in 1920.

Pierce Purcell was fine golf player and was captain of Portmarnock GC in 1925 and again in 1937. He served as president of the Golfing Union of Ireland and on the Royal and Ancient championship committee. He resided at "Ashton", Killiney, county Dublin.

He died at Jervis Street hospital in Dublin on 14 January 1968, following a fracture of his left femur, predeceased by his wife.

JACKSON, James Thomas (1868-1938), civil engineer, was born at Ballyoran, county Offaly on 17 May 1868, the son of William Jackson and Anna Ashbury, Methodist farmers.

Jackson was educated at Santry School in Dublin and entered Trinity College Dublin on 28 June 1890 with a Mathematical Sizarship (scholarship). He went on to gain a Senior Moderatorship and Large Gold Medal in Experimental Science and a Junior Moderatorship and Silver Medal in Mathematics, and was awarded his BA degree by the University of Dublin in 1894. He then studied civil engineering and received his BAI degree with special certificates (honours) in 1896. He was later awarded the degrees of MA (1897) and MAI (1900).

From October 1896 to September 1897, he acted as works manager for the Natural Colour Photo Co. (the Joly process of colour photography). From November 1897 he was employed by Luke Livingston Macassey as an engineering assistant on the Belfast waterworks, where he was involved in the Mourne Scheme, preparing contract drawings for the engines and boilers for Oldpark pumping station, the Ballysillan reservoirs, and the conduit from the Silent Valley, including being resident engineer on the three-mile long Knockbreckan tunnel. In May 1899, Jackson sailed for Siam to take up the headship of the Royal Ayuthia Survey School and acted as officer-in-charge of a section of the Ayuthia Survey.

Returning to Ireland in 1901, he joined the staff of the School of Engineering at Trinity College Dublin as a lecturer and assistant to the professor of natural and experimental philosophy, William Edward Thrift, and from 1907 as second assistant to the professor of civil engineering, Thomas Alexander. He became a close colleague of Walter Elsworthy Lilly, who had been appointed lecturer in mechanical engineering. Jackson became interested in keeping records of former engineering graduates and was responsible for the first *Record of the School of Engineering* (1909), thus laying the foundations for a publication that survived in various editions until 1993.

On 7 June 1899, Jackson married Elizabeth Louise, daughter of Lewis Damer Mulhall, a commission agent from county Laois. They had three sons, Desmond Ashbury (1902-1913), Guy Damer Jackson (1905-1984) and Paul Alexander (1913-1990), the last two becoming civil engineers like their father. The family lived at 37 York Road in Rathmines, Dublin.

Jackson was elected AssocICEI in 1896, transferred to AMICEI in 1899, and to MICEI in 1907. He acted for a few years as Hon.Secretary, before becoming president in 1924. He was elected AMICE in 1899.

Jackson died in Dublin, following an operation for prostate cancer, on 22 November 1938, and is buried in Mount Jerome cemetery.

HASSARD, Arthur (1875-1949), civil engineer, was born at Clones, county Monaghan on 15 July 1874, the son of Robert Hassard and Frances Reed from county Down.

He was educated at Corrig School, Kingstown (Dun Laoghaire) and entered Trinity College Dublin in 1895, where he studied civil engineering. He received the degrees of BA and BAI in 1897 from the University of Dublin, and an MAI in 1907. He received his early training under Berkeley Deane Wise, the chief engineer of the Belfast & Northern Counties Railway, before acting as an assistant engineer to Wise for a further year.

Towards the end of 1899, Hassard became personal assistant to William Collen, County Surveyor of Dublin, but in 1900 a combined engineering staff was formed to carry out engineering works for the Congested Districts Board and for the Department of Agriculture and Technical Instruction for Ireland. In January 1901, Hassard became chief assistant to the Chief Engineer,

Charles Deane Oliver. In April 1905, the entire staff was transferred to the Department of Agriculture and Technical Instruction for Ireland. Surveys, preparation of plans and designs for, and supervision of many harbours and piers along the northern and western seaboards were carried out by the department between 1905 and 1916, including piers at Tramore, Poulduff, Rathlin, Annagassan, and many other smaller piers, together with programmes of harbour dredging. Structures completed included a thirteen-arch masonry bridge at the exotic-sounding location of Muckinaghtetrahetrahaulia in county Galway. In February 1916, Oliver was seconded for special work at Kynock-Arklow Ltd. in connection with their war contracts, and Hassard was appointed Acting Chief Engineer of the department, with sole responsibility for the marine and fishery engineering work. His office was at 2 Kildare Place in Dublin, and later at 24 Kildare Street (1927) and at Hume Street (1928).

On 17 June 1908, Hassard married Lillian Alice, daughter of William Harman from county Meath. The family resided at 59 Grosvenor Square, Rathmines, Dublin and later at "Churchville", on Church Avenue in Rathmines.

Hassard joined the ICEI as a student member in 1897, was elected MICEI in 1911, served on council, and was president in 1926. He was elected AMICE in 1900, and transferred to MICE in 1919. Hassard died at "Churchville", Church Avenue, Rathmines, Dublin on 21 January 1949.

DELAP, Alfred Dover (1871-1943), civil engineer, was born at Maghery in county Donegal on 17 June 1871, the third son of Canon Alexander Delap of Valentia, county Kerry, and Annie Goslett. Alfred was educated in Clonmel and entered Trinity College Dublin in June 1890. He gained his BA degree in 1893, and a BAI (engineering) degree with special certificates (honours) in 1895, both from the University of Dublin.

Following graduation, Delap was employed by Samuel Gordon Fraser, as his assistant on railway plans, and on plans for water and sewerage works, until 4 May 1897, when he was appointed engineer to the Southern State Gold Fields Development Company and left for America where he was employed for some months in mining and civil engineering work.

On returning to Ireland in 1898, Delap became assistant to Sir Benjamin Baker and Kennett Bayley, engineers to the Fishguard & Rosslare Railways & Harbour Company. The following year he was employed by them on the survey for the Cork & Fermoy Railway, on the survey for the river Suir bridge, and on the contract, survey and plans for the Waterford & Rosslare Railway.

In 1900 he completed surveys for the harbour at Rosslare and, in the latter part of 1901, was appointed resident engineer on behalf of the company for the works at Rosslare Harbour. The works consisted chiefly in widening and raising an existing pier, adding another 1,100ft to its length, and the construction of a steel rail viaduct to give access to the pier.

Following the completion of the works at Rosslare Harbour in 1912, Delap founded the consulting engineering firm of Delap & Waller in partnership with James Hardress de Warrenne Waller, a practice that was expanded greatly in later years. Their office was at 115 Grafton Street in the heart of Dublin and their telegraphic address was 'Reinforced, Dublin'.

In the period 1912-1915, he constructed Helvick Harbour in county Waterford for the Department of Agriculture. During his career as a consultant, he was responsible for the design of a large number of public works, including the expansion of Foynes Harbour on the Shannon estuary, the Deeps Bridge over the river Slaney, Mountgarrett Bridge over the river Barrow, and a number of other bridges in Kerry and Tyrone. He was recognised at the time as one of the country's leading experts in reinforced concrete design.

In later years, Delap became involved with a number of proposed schemes, including assisting Sir John Purser Griffith with proposals to harness the river Liffey for hydro power, and suggesting ways in which areas of Dublin Bay could be reclaimed as development land. He also worked on the development of tidal power.

On 4 October 1905, Alfred married Jane Ethel, daughter of Thomas Jefferies of Newbay, county Wexford. Their only child, Hugh Alexander Delap, followed in his father's footsteps, becoming in time Chief Engineer of the Board of Works in Dublin and a president of the Institution of Engineers of Ireland.

Delap was elected MICEI in 1913, was a member of council 1917-19, 1922-23, 1924-25, 1929-1943, vice-president 1926-27, and president 1927-28 and 1928-29, during which period he was largely responsible for the adoption of a standard scale of fees for members.

He was elected AMICE in 1897 and transferred to MICE in 1907.

Alfred Delap died at his residence, "Dangan", Carrickmines, county Dublin on 3 October 1943.

BUCKLEY, Michael James (1870-1951), civil engineer, was born at Walshestown near Newbridge, county Kildare on 1 October 1870, one of three sons of Peter Buckley, farmer, and Mary Kinsella. His brother, Donal O Buachalla, served as the last Governor-General of Ireland.

He was educated at Catholic University School and studied engineering at the Royal College of Science of Ireland, gaining his Associateship in 1892.

After graduation, he spent two years conducting surveys for the Land Commission and also some time with the Commissioners of Irish Lights. In 1894 he became an assistant to Spencer Harty in Dublin Corporation and in 1897 was appointed township engineer and surveyor to Drumcondra township, and continued in that post until that area was incorporated into the city of Dublin in 1900. He then became engineer to the City Council Improvement Committee and in 1906 was placed in charge of the city's main drainage scheme. In 1913 he was appointed borough surveyor and engineer in charge of the waterworks (then at Roundwood), in succession to John Gabriel O'Sullivan. He remained as city engineer and borough surveyor of Dublin until his retirement in 1925, when he was succeeded by Michael A Moynihan. From 1932 to 1934 Buckley was a member of the National Housing Board.

Elected MICEI in 1895, he served on council 1917-1924, was vice-president 1924-1928, and became president in 1929-30. He presented a paper to the ICEI in 1902 on *Steam Rolling of Roads*, for which his was awarded a Mullins Silver Medal. He was also elected AMICE in 1897, and was a member of the Association of Municipal & County Engineers.

On 31 July 1901, Buckley married Bride, second daughter of John K O'Connell of Finglas Road, Dublin. They had two sons, Padraig and Aonghus.

Michael Buckley, who resided at 19 St Helen's Road, Booterstown, county Dublin, died at St Laurence's (Richmond) Hospital in Dublin on 12 May 1951 and is buried in Glasnevin cemetery.

MALLAGH, Joseph (1873-1959), civil engineer, was born at Markethill, county Armagh on 20 August 1873, the son of Joseph Mallagh, a merchant, and Margretta Gray.

He was educated at Monaghan Collegiate School (1887-1890) and at Queen's College Galway, where he entered in 1892 to study civil engineering. He graduated in 1896 with the degrees of BA and BE with honours from the Royal University of Ireland. He received practical training until 1897 with Edward Townsend, and from 1897-98 with Alfred Dickinson Price, Town Engineer of Blackrock, county Dublin. Mallagh was responsible for the extension of waterworks at Bray and Greystones in county Wicklow. In 1899 he was appointed resident engineer for the Downpatrick waterworks improvement scheme under Hassard & Cowan of Newry.

For a period from 1901, he was Surveying Officer for rating purposes on the staff of the General Valuation Office (Ireland) in Dublin. In 1903 Mallagh became assistant to John Harris Hazlett Swiney, on the preparation of plans and estimates for Portadown and Banbridge waterworks and was subsequently in charge of the construction of Boyle waterworks in county Roscommon. In 1905, Mallagh was appointed Engineer to the Sligo Harbour Commissioners and designed and carried out schemes of harbour improvement and maintenance during his twelve years with the commissioners.

He married Bertha Sloan from county Roscommon in 1907 and is recorded as living at 14 Wine Street in Sligo. A son, Terence John Mallagh was born around 1910 and followed in his father's footsteps, becoming a civil engineer and in time taking over his father's practice.

In 1917, Mallagh became Chief Engineer to the Dublin Port & Docks Board and was thereafter responsible for the design and carrying out of all improvement work, including a deepwater wharf. He was also consultant to several other port authorities in Ireland. On retiring from Dublin Port in 1940, he set up as a consultant and was joined in 1949 by his son Terence John Mallagh to form Joseph Mallagh & Partners.

He was elected AMICE 5 December 1899 and transferred to MICE in 1931. He was elected MICEI 1909, served on council, was a vice-president 1927-30, and became president in 1930. Joseph Mallagh died at his residence, 39 Nutley Park, Dublin, on 1 November 1959.

GALLAGHER, Stephen Gerald (1871-1959), civil engineer, was born at Waterford on 27 July 1870, the son of Antrim-born Thomas Gallagher and Marianne Bible. At the age of ten, Stephen was living in Southampton, where his father was a civilian assistant in the Ordnance Survey HQ in the city. He was educated at Synge Street CBS, Blackrock College, and Queen's College Galway, where he studied civil engineering. He received a BE degree from the Royal University of Ireland in 1896. He commenced his engineering career as a contractor's engineer on the Blackrock & Kingstown main drainage (1895-1900), on the Dublin main drainage ((1896-1899) and on the Clontarf & Howth tramway (1899-1900). From April 1900, Gallagher was involved with a number of public works for Wicklow county council as assistant county surveyor and was appointed Surveyor & Engineer to the Rathdrum Board of Guardians and, in the same capacity, to the Rathdrum RDC. At the time of the 1901 census, he is recorded as living at 4 Lower Street in Rathdrum, he also resided at Corballis Castle in Rathdrum.

In May 1903, Gallagher sat and passed the qualifying examination held by the Civil Service Commissioners and in August of that year was appointed County Surveyor for county Wicklow, where he was to serve until his retirement in May 1938. In 1911, his family is recorded as living at 8 Wentworth Street in

Wicklow. During his time as county surveyor, Gallagher, like many of his fellow surveyors, had to deal with a deteriorating road network, the damage being occasioned by the increase in heavy commercial transport. In 1907 he tried to persuade the council to permit the acquisition of steamrollers and to introduce direct labour schemes, but it was to be another ten years before improvements works were commenced with sections of the Bray to Wicklow main road. Direct labour schemes were finally introduced in November 1920 and a large fleet of road machinery built up over the next five years, although direct labour was not applied to all road maintenance until the mid-1930s.

On 9 July 1900, Gallagher married Catherine Elizabeth, London-born daughter of Thomas Carroll, a merchant. A son, Maurice Barry (1903-1977) and three daughters are recorded in the 1911 census.

Gallagher was elected MICEI in 1903, and served on council, and was president in 1931-2.

He died at his residence "Wentworth", Novara Avenue, Bray, county Wicklow on 7 April 1959.

KETTLE, Laurence Joseph (1878-1960), electrical engineer, was born at Kilmore, county Dublin on 27 February 1878, the eldest son of twelve children of Andrew J Kettle, a progressive farmer, and Margaret McCourt of Finglas, county Dublin, who was nearly twenty years his younger. Andrew Kettle was a founder member and secretary of the Irish Land League, and adviser to and confidant of Charles Stewart Parnell. Laurence was a brother of Tom Kettle, poet, essayist and patriot, whose statue stands in St Stephens Green.

From 1888 to 1897, Laurence received his early education at the Christian Brothers O'Connell Schools in Richmond Street, Dublin and at Clongowes Wood College in county Kildare. Between 1897 and 1902, Kettle served a pupilage with Hugh Erat Harrison, MIEE, AMICE, during which time he attended practical and theoretical courses of instruction in electrical engineering at the Electrical Standardizing, Testing and Training Institution at Faraday House in Charing Cross in London. He studied on a two-year Maxwell Scholarship, graduating with first class honours, and gaining first place in the diploma examinations. Following graduation, he trained with Robert Stephenson & Co., at Newcastle-upon-Tyne, and then with Maschinen-fabrik Oerlikon and Société des Forces Électriques de la Goule in Switzerland

In 1902, he moved to St Omer in France to work for a year as an assistant engineer with a company manufacturing turbines, the following year moving back to Oerlikon to supervise the installation of 4,000 HP turbines for the Dublin electricity works at the Pigeon House. In 1904 he carried out the re-design and re-construction of steam plant in Dublin for the General Electric Co. The following year he was appointed chief assistant engineer for Oerlikon based in London.

In 1906 Kettle was appointed works superintendent engineer at the Dublin Corporation electricity works, working under the City Electrical Engineer, Mark Ruddle. In 1916 Kettle (known as 'Larry' to his intimates) became Deputy City Electrical Engineer and in 1918 succeeded Ruddle with the new title Engineer & Manager of the electricity department of Dublin Corporation.

When the undertaking was taken over by the Electricity Supply Board (ESB) on its establishment in 1929, Kettle became an adviser to the Board and in 1934 a full member of the Board until his retirement in 1950. The development of the Liffey hydroelectric schemes and the later generation of electricity from turf-burning power stations were largely due to his efforts. For over fifty years, Dr Kettle was keenly interested in problems of the national economy and wrote widely on the subject.

In 1934 he was appointed first Director of the Industrial Research Council and held that office until 1946. In that year the larger Institute for Industrial Research and Standards was established and Kettle became its first Chairman. He served as Director of the Industrial Credit Co. from its foundation in 1933 until 1947. He was elected MICEI in 1907 and served as president in 1932. On 5 January 1920, Kettle read a paper before the ICEI entitled 'Irish Coal and Coalfields' for which he was awarded a Mullins Silver Medal. Kettle was also a member of the IEE, having joined as a student member in 1899, elected an associate in 1902,

121

associate member in 1906 and a member in 1912. He was Chairman of the Irish Centre of the IEE (1926-28) and served as honorary treasurer 1933-34. He was a member of the Irish Committee of the IMechE and a fellow of the Institute of Fuel. He was a founder member of the World Power Conference and secretary of its Irish committee for many years. In 1938 he was awarded an honorary DSc by the National University of Ireland. The Kettle Premium award in the Institution was a bronze plaque, a replica of a bas-relief executed by Albert Power in 1933 showing Kettle's side-face.

Laurence Kettle and his brother Thomas M. Kettle were two of the sixteen members of the provisional committee responsible for forming the Irish Volunteers in 1913. Laurence was a member of the Irish Parliamentary Party and served as an 'honourable secretary' in the movement. He was author of *Material for Victory: The Memoirs of Andrew J. Kettle, Right Hand Man to Charles Stewart Parnell* (Dublin, 1958). A portrait of Laurence Kettle by Sean Keating was presented in 1939 to the Hugh Lane Gallery in Dublin.

Laurence Kettle died at 46 Cowper Road, Rathmines, Dublin on 27 August 1960. The funeral was to Swords.

O'DWYER, Nicholas (1895-1957), civil engineer, was born in the village of Rahen, near Croom, county Limerick, on 20 February 1895, the eldest son of John O'Dwyer, farmer, and Bridget Hayes. Nicholas was educated at St Colman's College, Fermoy, county Cork (1907-13) and in 1913 entered University College Dublin to study engineering, graduating BE from the National University of Ireland in 1916. He then undertook research at UCD into the strength of concrete mixes and acted as a demonstrator in the department under professor Pierce Purcell. Between 1917 and 1919, he was employed building creameries by Wall & Forde, contractors, of Bruff, county Limerick..

During the summer of 1919 he worked on the Department of Agriculture's Tillage Survey and for the academic year of 1919-20 taught building construction at Limerick Technical Institute. He was then caught up in the independence struggle, in which he played an active role as brigade engineer and battalion commander in the East Limerick Brigade. On 8 February 1921 he was appointed an engineering inspector in the pre-Truce Department of Local Government, and on 1 July 1922 joined the staff of the new Ministry of Local Government with the same rank. He was promoted to the post of assistant chief inspector in 1928. In 1930 he was appointed commissioner to administer the affairs of Kerry County Council, which had been dissolved on account of irregularities in its procedures. At about the same time the Public Health Section of the League of Nations selected him to be a member of the committee of experts formed to prepare a report for the European Conference on Rural Hygiene in 1931.

In July 1931 he was promoted to the position of chief inspector for the Ministry of Local Government, but within a year he had resigned in order to establish a private engineering consultancy in Dublin; he was also retained by a number of local authorities in various parts of the country to carry out drainage, water supply and housing schemes. Water supply and sewerage schemes were to be his principal business for the rest of his life. His office was a thriving one, and many young engineers gained their first engineering experience in it.

He was also responsible for a number of new bridges, and the No.2 graving dock at Dublin Port.

His leisure interests included hunting and show jumping, in which he himself competed. He was master of the South Dublin Harriers for nine years in succession.

On 28 June 1922, O'Dwyer married Catherine Teresa Hayes, daughter of John Hayes (deceased), a farmer, from Bruff in county Limerick. One of his sons, Sean, also became an engineer.

O'Dwyer was elected MICEI in 1924; was a member of council 1927-1934 and 1936 until his death, and served as president, 1934-36. He was the youngest president to hold office and the first engineering graduate from UCD. His term of office coincided with the Institution's centenary celebrations. He was also president of the Engineers' Association 1939-40 and of the Association of Consulting Engineers of Ireland in 1958. O'Dwyer died at 6 Burlington Road, Dublin on 1 October 1957 and was buried in Deansgrange cemetery.

RISHWORTH, Frank Sharman (1876-1960), civil engineer, was born at Ballymoate Lodge, Tuam, county Galway on 5 April 1876, the son of John Henry Rishworth, gentleman farmer, and Hannah Sharman. He was educated at Ranelagh School, Athlone, at Santry School, Dublin and at Queen's College Galway, where he studied civil engineering from 1894 to 1898. He received a BA degree from the Royal University of Ireland in 1897, and a BE in 1898, taking first place at the examinations.

From January 1899 to 1900 he was assistant to William Haines Hutchinson on the construction of the Leen Valley extension of the Great Northern Railway (GNR) in Nottinghamshire, and from May 1900 until January 1901 an assistant in the construction department of the GNR. He then moved to the engineer's office of the GNR at Kings Cross in London where he was engaged in various aspects of railway construction work, including sections of the London underground system.

In 1902 Rishworth joined the Egyptian Civil Service and trained engineers for irrigation, railway and other public works at the Polytechnic School of Engineering at Ghizeh in Cairo. He designed and supervised the erection of a new engineering building on the campus. He was engaged in experimental work in connection with the Aswan dam and other irrigation structures. He retired from the Egyptian Civil Service in 1910 to take up an appointment as professor of civil engineering at University College Galway (UCG). He lived at 'Gort Ard'. Rockbarton in Galway.

In 1925, Professor Rishworth obtained leave of absence from UCG to become chief engineer on the Shannon hydroelectric scheme. A visionary and innovator, his influence on Thomas McLaughlin (the originator of the scheme) was seen as one of the prime motivating factors in the scheme. Rishworth was responsible to the government for the preparation of the contract documents, and the supervision of the civil construction works. Following its completion in late 1929, he returned to UCG, but continued to consult with the government and local authorities on a variety of issues. His report of 1932 paved the way for the installation of an extra turbine at Ardnacrusha in 1934. He moved to Dublin and commuted to Galway during the week. He was particularly active in the areas of water supply, sewerage schemes, and traffic engineering. He retired from UCG in 1946. He was identified with a number of public projects and served on the committees of professional and cultural associations as well as on the Drainage Commission, the Industrial Research Council, and the senate of the university.

On 17 July 1907, in Bingley Parish Church, Yorkshire, Rishworth married Mary Ann, daughter of the late William Beecroft, a manufacturer.

Rishworth was elected AMICEI in 1902, transferred to MICEI in 1912, was a member of council, and served as president 1936-38. He helped to establish the Institution's Benevolent Fund. He received an honorary degree of MAI from the University of Dublin in 1932 and in 1942 an honorary LL.D from the National University of Ireland.

Professor Rishworth died on 31 March 1960 at his residence, 29 Leeson Park, Dublin and is buried in Enniskerry churchyard, county Wicklow.

RYAN, Joseph Albert (1888-1971), civil engineer, was born in Dublin at 14 Sandford Road, Dublin on 18 August 1888, a son of William Alfred Ryan, a barrister, and Christine Byrne. He was educated at the Marist Fathers School in Leeson Street and from 1901 to 1905 at St Vincent's College Castleknock.

He commenced his engineering career by becoming for two years an apprentice with the Kerry Electricity Supply Co. at Killarney. He entered Trinity College Dublin in May 1907 to study civil engineering and received the degrees of BA (1910) and BAI (1911) from the University of Dublin. Following graduation he became an assistant surveyor with Pembroke UDC before sailing for Canada. He was employed between 1913 and 1914 on construction work (as an instrument man) by the Grand Trunk Pacific and Pacific Great Eastern railways in British Columbia.

Returning to Ireland, Ryan was appointed in April 1915 to the position of county surveyor for Queen's county (Laois), having passed the qualifying examination the previous year. It is reckoned that he was the youngest ever appointment to a county surveyorship. Like many of his contemporaries, following the Civil War, he had to deal with the substantial damage to the roads and bridges in the county.

In July 1924, Ryan was appointed County Surveyor for Dublin. His term of office was notable for the major programmes of road development, housing and sanitary services. He undertook a traffic census in 1924 on the main roads and experimented with different road surfacing materials on short sections of the Naas road near Newlands Cross, the results being presented in January 1926 in a paper to the ICEI.

Ryan retired in July 1953, but worked thereafter for a number of years as sales manager of the Dublin firm of Moracrete Ltd.

On 26 July 1921, Ryan married Mary Josephine Rock, daughter of Christopher Rock. They had three sons, one becoming a doctor, the others prominent businessmen in the city.

Ryan was elected MICEI in 1920, was Hon.Secretary for a number of years, vice-president 1934-38, and served as president 1938-40. He was also chairman of the Irish District of the Association of Municipal and County Engineers 1925-6.

Ryan died in Dublin on 16 September 1971 and is buried in Deansgrange cemetery.

WALSH, Henry Nicholas (1892-1958), civil engineer, was born at 2 Janeville, Cork on 10 September 1892, the son of Patrick Walsh, a draper from county Kilkenny, and Mary Ellen Jones. Henry was educated at the Christian Brothers Schools, Our Lady's Mount and North Monastery (1899-1911) and at University College Cork (UCC), where he studied civil engineering. Having graduated BE from the National University of Ireland in 1914 with first class honours, Walsh worked first as a demonstrator at UCC, where he undertook research on timber columns, and then was a Beit Research Fellow in the civil engineering department at Imperial College London. He continued his structural research for the next two years obtaining the Diploma (DIC) in 1917. In the same year, he received the degree of ME from the NUI and renewed his association with UCC when he was appointed a lecturer in the department. From 1919 to 1920, he acted for Professor C.W.L.Alexander during the latter's illness and subsequent death.

In 1921, Professor Walsh was appointed to the chair of civil engineering at UCC and, for the next 36 years guided many students who, later in life, secured eminence in their profession. He lived at "Rossalia", Shanakiel Road, in Cork. Walsh published widely on the subject of concrete, the research carried out with a fellow lecturer, P.Coffey, resulting in a considerable

contribution to the knowledge and development of the subject. His book *How to Make Good Concrete* became a standard text. In 1924 he was awarded a Mullins Gold Medal by the ICEI for his paper entitled 'The design of concrete mixes' and again in 1943 for a paper on 'Aggregate grading and concrete quality'.
Walsh was appointed to many government commissions and had a special interest in the development of the country's natural resources, in particular the utilisation of water power. He was an early advocate of what became the Shannon hydroelectric scheme and was an automatic choice to serve on the 'Power Committee of the Dail Eireann Commission of Engineering into the Resources of Ireland'. During 1938-40 he served on the important 'Drainage Commission' and was a member of the 'Commission for Industrial Research and Standards' from its inception. Professor Walsh was very active as an engineering consultant, many water and wastewater schemes in Munster benefiting from his expertise.
Following the death of his first wife, Walsh, on 25 April 1928, married May Cecilia Mary O'Reilly, daughter of John O'Reilly, a Dublin manufacturer.
Professor Walsh was elected AMICEI in 1918 and MICEI the following year. He served on council, was a vice-president 1938-40 and president 1940-42. He was president of the Engineers Association in 1948. He was elected AMICE in 1937 and transferred to MICE in 1958. He served on the governing body of UCC from 1925 to 1957 and for several years on the Senate of NUI. He was also Dean of the Faculty of Engineering.
Having retired in 1957, and living at Riverview, Sundays Well, Cork, professor Walsh died of heart disease at CBS North Monastery on 27 November 1958.

MONAGHAN, Thomas Joseph (1887-1971), telecommunications engineer, was born at Toneen, Granard, county Longford on 10 April 1887, the son of John Monaghan, a farmer, and Kate Mary Carrigy.
From 1899 to 1902 he was educated at Xaverian College, Mayfield, Sussex and from 1902 to 1906 at South Western Polytechnic (University of London), where he gained a BSc (Eng) London with first class honours. Following graduation, he received practical training variously at the Royal Small Arms factory at Enfield, Yarrow & Co. at Poplar, and Simpson-Worthington at Newark until 1909.
Monaghan joined the General Post Office (GPO) as an assistant engineer at HQ in London. He was stationed in Belfast (1910-12), then in Dublin (1912-14) and then in Cork (1914-19), during which time he was in charge of works and maintenance of telegraphs, telephones, and wireless.
In 1919 he was transferred to the London HQ of the GPO on special wireless work and in 1921 became officer-in-charge of the Leafield Wireless Station in Oxfordshire responsible for plant erection and the running of what was then one of the principal wireless stations in the world.
Returning to Ireland in 1923, he became staff engineer at Department of Posts & Telegraphs in Dublin responsible under the engineer-in-chief for a variety of aspects of the work of the department, including the initiation of the Irish broadcasting service. He became engineer-in-chief in 1930 and moved to 15 Clarinda Park North, Dun Laoghaire. He retired in April 1952.
A chartered engineer, Monaghan was elected MICEI in December 1929, served on council, was a vice-president and served as honorary secretary, becoming president in 1942. He was also a member of the IEE and served as chairman of its Irish branch.
Monaghan married in 1909, a daughter Judith emigrating to the USA. He died at 18 Summerhill Road, Dun Laoghaire on 10 July 1971.

COURTNEY, Thaddeus Cornelius (1894-1961), civil engineer, was born in Cork on 13 December 1894, the elder son of Timothy Courtney, a clerk and former member of the Royal Irish Constabulary, and Ellen Shea, both from county Kerry.

He was educated in Cork, first at CBS North Monastery and then at Presentation College, before entering University College Cork in 1913 to study civil engineering. He received his BE degree in 1916 and an ME degree in 1932, both from the National University of Ireland.

Following graduation, Courtney trained for a year as an assistant with the Cork, Bandon & South Coast Railway and then with Henry Ford & Sons on the design and construction of their new factory at Marina in Cork. Moving in 1918 to Harland & Wolff in Belfast, he worked on a number of projects connected with the expansion of the shipyards. In 1920 he returned to Fords for a further two years for the completion of the works, until, in 1922 he joined the newly formed Irish army to assist with the organisation of the Corps of Engineers, retiring in 1925 with the rank of Major.

In 1925, Courtney was appointed an engineering inspector (roads) at the Ministry of Local Government and Public Health, where he was engaged mainly in the supervision of road construction being undertaken by local authorities. From 1930 to 1934 he was county surveyor for Tipperary (North Riding), town surveyor of Thurles Urban District Council (UDC) and also of Templemore UDC, responsible for the maintenance and construction of roads and bridges.

Returning to the ministry, he was appointed Chief Engineering Adviser, responsible for advising the minister on all matters relating to public health works, housing, roads, and hospital construction. He lived at 36 Leeson Park in Dublin.

He was called upon to serve on several government commissions, inquiries, and special projects. In particular he served on the Greater Dublin Tribunal and the 1939 Housing Commission. During WW2 ("The Emergency"), he was chairman of the Turf Executive responsible for the production of fuel, and for ten years as part-time railway inspecting officer.

In 1949 Courtney changed from being responsible for directing the large investment (£55m) in roads to the modernisation of the railways, As chairman of Córas Iompair Éireann, he was successful in guiding the organisation in its programme of dieselisation. He retired in 1958 just as the last of the steam locomotives were being phased out.

On 5 August 1925, Courtney married Letitia Josephine, daughter of James Fitzsimons, a clerk, from Downpatrick, county Down.

He was elected AMICEI in 1919 and transferred to MICEI 1928. He served on council 1935-42 and 1944-61, was vice-president 1942-43 and president 1943-44. He was elected AMICE in 1936. He was president of the Engineers Association in 1953. With George Howden, he was responsible for forming and fostering the Irish Branch of the Institute of Transport, of which he became chairman. 'Ted' Courtney died at Waterville, county Kerry, on 5 August 1961.

RAFTERY, Patrick Joseph (c1886-1962), municipal engineer, was born at Corofin, county Galway about 1886, a son of Patrick Raftery, a merchant. He was educated at Mungret College, Limerick and at Queen's College Galway, later transferring to University College Dublin. He received his BE degree from the National University of Ireland in 1916.

In 1917 Raftery was appointed assistant county surveyor in Galway, and held that post until he joined the Department of Local Government in 1922 as an engineering inspector. He became one of a team that planned and helped accomplish the massive road restoration programme of the Irish Free State, including the £2 million road improvement programme of 1926. While in the department, he took part in many enquiries into the acquisition of property and the dissolution of public bodies. He retired as Senior Engineering Inspector in 1953.

On 10 January 1917, Raftery married Margaret, daughter of Thomas Halligan, a county Meath merchant.

Raftery was elected AMICEI in 1917 and transferred to MICEI in 1921. He served on the council 1926-27 and 1929-44, was Hon Treasurer 1929-47, and founder member of the Benevolent Fund. He was vice-president 1943-44 and became president 1944-45. Raftery was a member (1921) and vice-president (1961) of the Institution of Municipal Engineers, and was actively involved with the amalgamation of that institution in 1921 with the Association of Municipal & County Engineers, of which he was also a member. He was elected a fellow of the Royal Sanitary Institution in 1925. A keen antiquarian, he was founder member and president of the Old Dublin Society as well as being a member of the Georgian Society.

He died at 64 Upper Leeson Street in Dublin on 23 January 1962, survived by his wife and son, Patrick, who later became president of the ICEI. Patrick Joseph Raftery is buried in Mount Jerome cemetery.

CHANCE, Norman Albert Mary (1887-1963), municipal engineer, was born at Ivy Dene, Richmond, Belfast on 10 May 1887, the eldest of eight sons of Sir Arthur Gerard Chance, surgeon, of 15 Westland Row, Dublin, and his first wife, Martha Josephine Rooney. As a young man Norman lived at 90 Merrion Square West in Dublin.

He was educated at Clongowes College from 1898 until 1904, when he entered Trinity College Dublin to study civil engineering. He received the degrees of BA and BAI in 1907 from the University of Dublin.

The first four years of his career was spent with Messrs Shackleton of Carlow investigating small hydroelectric installations on the river Barrow. In 1911 he became a junior assistant engineer under the Dublin City Engineer, John G O'Sullivan and in 1914 a senior assistant engineer with Michael J Buckley. He enlisted in the Field Squadron Royal Engineers in 1916 and served until 1919, dealing with the preparation of road schemes and reports on bridges for the Roads & Bridges Department of the War Office. He retired with the rank of Major.

He joined Dublin Corporation in 1919 as Assistant Engineer in the City Engineer's Department in charge of the Paving Department. In 1936 Chance was appointed Dublin Borough Surveyor and Waterworks Engineer with responsibility for all engineering services. He retired at the end of March 1950. During his time with Dublin Corporation, major works completed included the Liffey water supply scheme, the Nose of Howth sewage outfall tunnel linked to the North East Dublin Drainage scheme, and the widening and regrading of a number of bridges over the Grand Canal.

He was elected MICEI in 1929, served on council 1936-7, 1938-44, 1946-61, was a vice-president 1944-45, and served as president 1945-6. He was president of the Engineers Association in 1951 and also served as chairman of the Irish branch of the Association of Municipal & County Engineers.

Norman Chance died at 62 Wellington Road, Dublin on 5 September 1963 and is buried in Glasnevin cemetery.

PURSER, John (1884-1967), civil engineer, was born at Parsonstown (now Birr), county Offaly on 1 November 1884, the only son of Alfred Purser, Chief Inspector of National Schools, and Ellen Hildebrand. Their three daughters all became teachers. A member of a Church of Ireland Dublin family, which had already produced five Trinity professors, John was educated at the High School in Dublin (1893-1902) and then entered Trinity College Dublin to study civil engineering. In 1906 he received the degree of BA and BAI the following year from the University of Dublin and proceeded to the University of Birmingham where he took an MSc degree in July 1910. He was later to receive the degrees of MA (1933) and MAI (1936) from his alma mater. He held the post of Lecturer in Civil Engineering at Birmingham for six years before moving to London to become Assistant Professor of Civil Engineering at the newly founded City & Guilds Engineering College (C&GEC). He resided at 56 Riggindale Road in Streatham.

At the outbreak of WW1, Purser joined the Royal Navy, serving for a time in coastal motorboats before becoming a research officer in the Mine Sweeping School on HMS Vernon at Portsmouth naval dockyard, where he spent two and a half years on research into the design and performance of paravanes (knife-bearing torpedo-like devices for cutting mines adrift). In 1919 he returned to the C&GEC in London to continue his academic career. From 1928 until 1933, Purser also held a Readership in Civil Engineering from the University of London.

From 1926 to 1933, he served as Secretary to the ICE Committee on the Deterioration of Structures of Timber, Metal, and Concrete exposed to the action of sea-water.

Following the sudden death of David Clark in 1933, Purser was offered the Chair of Civil Engineering (1842) at Trinity College Dublin. He came to Dublin at a time when the School of Engineering was in need of modernization and this applied to the curriculum as much as to the equipment and buildings. His main interest was in hydraulics and his model work for the Erne hydroelectric scheme (with Gerald FitzGibbon) was a major contribution towards its design.

During Purser's 24-year tenure of the Chair, the annual intake of students rose from 10 to 40, the Engineering School was partially re-equipped and, just before his retirement from the Chair in 1957, permission was granted to extend the engineering course from three to four years.

On 14 April 1914, at the Church of St Patrick, Kilcouriola, county Antrim, John Purser married Elspeth Marjorie Gordon, daughter of William Stuart, a civil engineer, of Mount Earl, Ballymena. John's aunt, Sarah Purser, was a noted artist.

Purser was elected MICEI in 1935, served as Hon. Secretary (1938-1943), was a member of council 1947-61, was a vice-president 1944-45 and served as president in 1946. He was elected AMICE in 1919 and served on the ICE Sea Action Committee (1926-33). A commissioner of Irish Lights (and at one time chairman), he maintained his maritime interests and gave valuable advice when the Kish Bank lighthouse was under construction.

He died at home in Dublin on 29 August 1967 and was buried in Calary churchyard in county Wicklow.

MacDONALD, Joseph (1893-1983), civil engineer, was born at The Mall, Waterford on 15 December 1893, the son of David MacDonald, baker and brewer, and his wife Anna Mary.

His early education was at Mount Melleray Abbey, Cappoquin, county Waterford. Between 1910 and 1913 he attended the Municipal Technical Institute in Belfast before entering University College Galway (UCG). He graduated with an honours BE degree from the National University of Ireland in 1917 and also gained a BSc (Engineering) from the University of London.

Following graduation, MacDonald commenced his career as an assistant engineer (1918-1920) under Joseph Mallagh, Chief Engineer, to the Dublin Port & Docks Board. He then spent a year with Sir William Arrol & Co. of Glasgow as a structural designer.

In 1925, Frank Rishworth, his former professor at UCG asked him to join his team of Irish engineers set up under the Department of Industry and Commerce to act for the government on the Shannon Scheme. He took a major part in checking the design of the civil engineering works and negotiating payments for these with the civil works contractors Siemens Bauunion.

In 1931, he was appointed to the newly created position of Chief Engineer (Civil Works Dept) of the Electricity Supply Board (ESB), with offices at 40 Merrion Square in Dublin. He developed the department to provide the civil engineering base for the rapidly expanding ESB. He organised the reconstruction of the old Pigeon House power station, the final Shannon storage works at Lough Allen, and the Liffey hydroelectric stations at Poulaphuca and Golden Falls. Following WW2, work started on the Erne and Leixslip hydroelectric schemes and the first peat-fired station was built at Portarlington. 'Mac' retired in December 1960, but continued as a consultant and arbitrator for many years thereafter.

MacDonald was elected AMICEI in 1920 and transferred to MICEI in 1936, served on the council 1937-64, was a vice-president, and served as president 1947-48. He was elected AMICE in 1926.

He died of stomach cancer at 7 Airfield Park, Donnybrook, Dublin on 10 October 1983, predeceased by his wife.

CANDY, Joseph Phelan (1895-1960), civil engineer, was born at Derry on 7 July 1895, the son of James Candy, stationmaster, and Mary Catherine Gribbon. Joseph was educated at St Columb's College in Derry (1908-1914), and studied civil engineering at Queen's University, Belfast (1914-1917), receiving a BSc degree in 1917 and an MSc in 1922.

Following graduation, he went to England to receive practical training under John Miller, Chief Civil Engineer of the Great Eastern Railway (GER). During the final months of WW1 he served with the railway construction branch of the Royal Engineers. In 1921 he became assistant docks engineer for the GER at Lowestoft harbour, moving in 1924 to become personal assistant to the chief engineer at Liverpool Street. He left the GER in 1926 and spent the next five years with the Public Works Department in Burma, mainly on irrigation works. He then became chief engineer and manager in the Near East for the Tilbury Contracting & Dredging Company, carrying out major works in Egypt and Palestine, including works at the port of Haifa.

In 1934 he was appointed Chief Engineer at the Office of Public Works (OPW) a position he occupied until his death shortly before he was due to take retirement. With his wide knowledge of public works organisation and a specialist knowledge of river drainage, earth moving, dredging, coastal and harbour problems, he was well-equipped to guide the engineering branch of the OPW during a period of expansion, which included a major programme of arterial drainage recommended by the 1938-40 Drainage Commission, of which he was a member. The recommendations were embodied in the Arterial

Drainage Act of 1945. Candy oversaw the completion of the first four catchment drainage schemes (Brosna, Glyde and Dee, Feale, and Nenagh). In connection with this work, he organised the collection of valuable hydrometric data. As chief engineer, he also had responsibility for the maintenance of the government harbours at Dun Laoghaire, Howth and Dunmore East and the construction and maintenance of many of the fishery harbours. The construction of the first grass runways and other works at airports at Dublin (Collinstown) and Shannon also came under his remit.

Some time before 1932, Candy married Agnes O'Kane from Buncrana, county Donegal.

Candy was elected MICEI in 1935, served as a council member 1937-60, was vice-president 1946-7, and became president in 1948. He was also elected AMICE in 1921 and transferred to MICE in 1931. Candy died at Dublin on 13 March 1960.

HOGAN, Michael Anthony (1898-1971), civil engineer, was born at 23 Dame Street, Dublin on 14 June 1898, the eldest son of Thomas Joseph Hogan, a grocer and licensed vintner, and his wife Joanna Mary Hogan.

His parents had moved to 27 Cowper Road in Rathmines by the time Michael began attending the Catholic University School in Lower Leeson Street (1907-1915). He entered University College Dublin (UCD) in 1915 to study civil engineering. In 1918 he gained a first class honours BE degree from the National University of Ireland (NUI) and was awarded a Pierce Malone Scholarship enabling him to carry out research into the raw materials available in the Dublin district for the making of concrete. In 1919, he was awarded an MSc in Geology.

In 1920, with a Beit Fellowship, he began two years post-graduate research at Imperial College, London, carrying out river gauging and flood prediction work for the investigations of the River Severn. He received the Diploma of Imperial College in 1921 and a PhD degree the following year from the NUI. In 1929 was awarded a DSc (Eng.) by the University of London for his published works. He carried out investigations with Professor Purcell in 1922 of the upper river Liffey, with a view to hydroelectric power development. He was also involved with him on the mechanical problems of peat winning machinery. In 1937 he received a Telford Premium for a paper to the ICE on the subject of his earlier River Severn investigations. In 1924, Hogan spent some time with Sir William Arroll & Co. of Glasgow on bridge design, including Waterloo Bridge in London. He next went to Siemens Bauunion in Berlin to work on the civil engineering design of the Shannon Scheme, including a tour of duty on site at Ardnacrusha. In 1926 he joined the Safety in Mines Research Board as a senior assistant to Professor Stephen Dixon of Imperial College, and remained with the Board until 1938. During this period, he carried out investigations into both the civil and mechanical engineering aspects of mine safety. He was made principal assistant in 1930, and on Professor Dixon's retirement in 1933 was placed in executive charge of all the Board's research work at Imperial College. In 1939, Michael Hogan returned to Ireland, having been appointed Professor of Mechanical Engineering at UCD, a post he was to hold for the next sixteen years. This appointment involved him, not only in the development of the university department, but also in the application of technology to many of the problems that arose during the 'Emergency' (WW2). In 1941, Professor Hogan was appointed a member of the Engineering Scientific Research Bureau, a body whose main function was the production of materials that were then in short supply in Ireland and not readily imported. He was also appointed a director of Mianrai Teoranta, and was subsequently chairman, and from then until 1956, was closely associated with mining exploration for coal and minerals in the country. He became a prominent member of the Institute for Industrial Research and Standards, the body responsible for publishing statutory Irish standards. In addition to his academic work, Dr Hogan was deeply concerned with the development of Ireland's natural resources and thus served the community on two fronts, educational and economic.

In 1954, Michael Hogan succeeded Pierce Francis Purcell as Professor of Civil Engineering and Dean of the Faculty of Engineering and Architecture at UCD and was responsible for an expansion in both undergraduate and postgraduate activities. Michael retired in 1968, but became deeply involved in the development of the new campus of his alma mater at Belfield in the south city, serving as chairman of the

Building Committee. In appreciation of his services to his country and to his college, the National University of Ireland conferred on him in 1969 the honorary degree of DSc.

On 21 July 1931, in St John's, Clontarf, Michael Hogan married Violet Mignon, a pensions officer, and daughter of the late Thomas Cannon, Superintendent, GPO.

Professor Hogan was elected MICEI in 1939, served on council and for a period as Hon.Secretary, and was elected president in 1949. He was elected AMICE in 1923 and transferred to MICE in 1936. He was also elected MIMechE in 1938 and a member of the RIA in 1939. He died at his home in Killiney, county Dublin on 6 October 1971.

McLAUGHLIN, Thomas Anthony (1896-1971), electrical engineer, was born at Dublin Road, Drogheda on 14 June 1896, the second son of Peter McLaughlin, a Customs & Excise officer, and Sarah McCabe. By the time of the 1911 census, there were thirteen children in the family, including twin boys.

Thomas was educated at the Christian Brothers School, Synge Street in Dublin (1908-1913) and at University College Dublin, where he studied mathematical physics, obtaining the degrees of BSc (1916) and MSc in Experimental Physics (1918) from the National University of Ireland (NUI). He then obtained a position as an assistant lecturer in the physics department of University College Galway (UCG). Whilst there he found time to complete studies for a BE in Electrical Engineering, the degree being awarded by the NUI in 1922. About the same time he completed his research into the behaviour of gas bubbles in a fluid subjected to the action of an electric field, for which in 1921 he was awarded a PhD degree by the NUI.

Before his PhD was conferred, McLaughlin went to Berlin to join the firm of Siemens-Schukertwerke as the first in a long line of Irish engineering trainees. During his time in Berlin he developed further his idea for the harnessing of the river Shannon to provide electricity, the idea having been sown in his mind whilst at UCG. In 1924 he was appointed manager of the firm's Dublin office and was able to persuade the Irish Free State government to accept a plan put forward by Siemens for the supply of electricity all over the country from a single hydro-power station (at Ardnacrusha on the Shannon) and a high-tension distribution system. An alternative scheme based on the river Liffey, and nearer to the majority of the consumers, was supported by many, including Sir John Purser Griffith and Dr Laurence Kettle, but the Shannon Scheme won the day and was completed by July 1929.

From 1924 McLaughlin served as Managing Director for Siemens in Ireland and transferred to the Electricity Supply Board (ESB) when it was formed in 1927. He resigned as MD in 1931 but subsequently was appointed one of three full-time directors of the ESB. Later, when the directorships ceased he engaged in a wide range of business activities, from civil engineering contracting to pharmaceuticals and oil distribution. He retired as a director of the ESB in 1957.

On 14 July 1926, McLaughlin married Olwen, daughter of Joseph O'Malley, a Limerick-based architect and engineer. They had three sons, Thomas, Desmond and Peter. The family lived at 19 Ailesbury Road (1954) and later at Simmonscourt Castle in Dublin.

McLaughlin was elected MICEI in 1929, and served as president in 1950-51. He had also served as president of the Engineers Association in 1946. He received an honorary doctorate from the NUI in 1940. Dr 'Tommy' McLaughlin died in hospital at Benidorm, Spain on 15 February 1971.

BLOOMER, William Ian Sidney (1900-1971), civil engineer, was born at Ballybrack, county Dublin on 26 June 1900, the son of William Henry Bloomer, a clerk of court from Bandon, county Cork, and Gertrude Quinn. Generally known as Ian, his father died when he was nine years of age.

Ian was educated at St Enda's, Rathfarnham, county Dublin, where he was taught by Patrick and Willie Pearse, and at Belvedere College, Dublin. He entered the engineering school at University College Dublin in 1918 and obtained a BE degree with honours from the National University of Ireland in 1921, followed by a BSc degree from the Royal College of Science in Ireland in 1922. After qualifying, Bloomer spent a year as a demonstrator in the engineering school before joining the newly formed Railway Protection, Repair and Maintenance Corps of the Irish Army. This involved him in the design and repair of bridges.

In 1923, Blooomer resigned his commission in the army to become resident engineer in charge of the reconstruction of the Taylorstown rail viaduct in county Wexford, four of the seven arches having been destroyed in the Civil War. He then went to London for a short period as assistant in the consulting engineering firm of J.F.Crowley & Partners, where, amongst other projects, he worked on plans for a proposed hydroelectric development on the river Liffey.

Returning to Ireland in 1924, Bloomer joined the Office of Public Works (OPW) as an assistant engineer on arterial drainage. However, from 1925 to 1930, he served with the Department of Industry and Commerce as a senior assistant engineer on the Shannon hydroelectric project, afterwards becoming a senior engineer in the newly formed Electricity Supply Board. In 1931 he went as assistant to the Town Surveyor of Newry, where he worked on flood relief works and a main drainage scheme. He also assisted Pierce Purcell at the early stages of the design of Kenmare Bridge. The following year he was appointed as an engineering inspector in the Department of Local Government and Public Health. In addition to his routine work, he sat on and conducted a large number of tribunals, commissions and enquiry boards. He acted as a consultant to the World Health Organisation. In February 1949, he succeeded Thaddeus Cornelius Courtney as Chief Engineering Adviser to the department, an office that he held until his retirement in 1965.

On 2 June 1936, Bloomer married Violet, a barrister, and daughter of Alfred Kimpton, a Dublin company director.

He was elected AMICEI in 1925, transferred to MICEI 1933, served on council, and became president in 1951. He was also elected AMICE in 1930. Ian Bloomer died at Killiney, county Dublin on 3 July 1971.

MURPHY Patrick Gerard (1900-1982), mechanical engineer, was born at 2 Lindsay Terrace, Glasnevin, Dublin on 19 May 1900, the son of John Murphy, commercial traveller, and Mary Ellen Kavanagh.

He was educated at O'Connell Schools in Dublin, where he was awarded a Dublin Corporation scholarship, and then in 1918 won an entrance scholarship to University College Dublin. He studied mechanical and electrical engineering and gained first place in the final examinations and received a first class honours BE degree from the National University of Ireland (NUI) in 1922. He was also awarded the ARScI.

Following graduation, Murphy worked for three years with civil engineering consultants J.F.Crowley & Partners of London, first in London and later in their Dublin office. For twelve months, from April 1923, Murphy acted as resident engineer for the government on the construction of the replacement Mallow rail viaduct and the repair of other viaducts damaged in the Civil War.

In 1925 he joined the firm of Siemens Schukertwerke in Berlin and was engaged on the design and construction of the Shannon Power Development scheme. On the formation of the Electricity Supply Board (ESB) on 11 August 1927, Murphy was one of the first two senior engineers to be recruited by Dr Thomas McLoughlin, ESB Managing Director. Murphy became Chief Design Engineer in 1932 in charge of all electrical and mechanical engineering construction work, including the reconstruction and modernisation of the Pigeon House generating station, and the extension to the Ardnacrusha station. In 1941 Murphy was awarded an ME degree by the NUI for a thesis on the design and operation of Kaplan hydraulic turbines.

Murphy was appointed Chief Engineer of the ESB in 1948. During his professional career with the ESB, 'P.G.', as he was known to his colleagues, was responsible for the building of the entire ESB system covering the distribution system, including rural electrification, the transmission system, and the generation of electricity from hydro, peat (turf), coal and oil. His outstanding contribution to the energy requirements of the State and its resulting industrial development – in the design and construction of the ESB system – was recognised and acknowledged by the NUI when it awarded him an honorary DSc in April 1965, following his retirement from the ESB. Murphy was a member of the Commission on Higher Education. He was also a member of the Atomic Energy Committee (1956-58).

Murphy was married around 1931, a daughter being born on 10 September, 1932. His residence was at "Sauglin", Hainault Road, Foxrock, county Dublin.

In 1925 he was elected AMICEI, later transferred to MICEI, served as a vice-president, and was president in 1952. He was also a member of Cumann na nInnealtóirí, serving as president in 1949-50. He was a member of the IEE and chairman of its Irish Centre. He was a member of the IMechE and served on its Republic of Ireland Committee.

Murphy died at St Joseph's nursing home, Kilcroney near Bray, county Wicklow on 31 December 1981.

NICHOLLS, Henry Nicholas (1889-1975), municipal engineer, was born at Derry on 28 December 1889, the only son of William Nicholls, a Shrewsbury-born national schools inspector and his wife Margaret from Tuam, county Galway.

His parents moved to Rathmines in Dublin and Henry was educated at Kingstown Grammar School, Kingstown and at Mountjoy School in Dublin. He entered Trinity College Dublin in 1907, was awarded scholarship, and became a moderator, receiving his BA degree from the University of Dublin in 1911. He then studied civil engineering and received his BAI degree from the university in 1913.

Following graduation, Nicholls joined the engineering staff of Dublin Corporation and trained under the City Engineer & Surveyor, Michael James Buckley, another president of the ICEI. As assistant engineer, he was in charge of the main drainage and sewers department, the work consisting mainly of the replacement of old sewers, and further development of the sewage pumping stations at Ringsend and Clontarf. From 1926 he was resident engineer on the construction of a service tunnel under the River Liffey between Ringsend and North Wall. By 1933 he was in charge of the city's sewers and main drainage.

Nicholls retired at the end of 1954 from the position of Corporation Sewers and Drainage Engineer and set up as a consultant at 42 Grafton Street, Dublin.

On 16 October 1918, Nicholls married Kathleen Maud, daughter of William Abel Holmes of Dublin.

Nicholls was elected AMICEI in 1921, transferred to MICEI 1923, served on council 1933-38, and 1954-61, was vice-president 1952-3 and elected president in 1953.

Nicholls died in the Meath hospital, Dublin, on 2 February 1975, predeceased by his wife.

FARRINGTON, Stephen William (1889-1965), municipal engineer, was born at 4 Waterloo Place, Cork on 6 May 1889, the son of Thomas Farrington, chemist, and Mary Emily Foreman. Stephen was educated at Cork Grammar School (1896-1906) and at University College Cork (UCC), graduating in 1910 with a BE in civil engineering from the National University of Ireland (NUI). He received the degree of ME from the NUI in 1933.

Farrington commenced his career in 1911 as an engineering assistant with Belfast Corporation working mainly on drainage schemes, including the Sydenham drainage scheme. In 1914, he became a works engineer with the contracting firm of H & J Martin, where he was engaged on the construction of a timber wharf, a road bridge crossing of the Great Northern railway at Finaghy, and various sewerage works.

In 1918, he served as an engineering assistant for the Admiralty on the construction of a naval station on Tory Island before moving to W J Campbell & Son on the contract for Aldergrove aerodrome. Following some bridge renewal work for the Belfast & County Down Railway, Farrington became Town Surveyor for Lisburn in county Antrim before, in 1924, being appointed City Engineer and Borough Surveyor of Cork and going to live at 7 Smithgrove Terrace in the city. He remained in post until his retirement in 1958. In Cork, he was responsible for many improvements in the road network in the city, for water and wastewater treatment facilities, and for the erection of the 160ft-span suspension pedestrian bridge across the river Lee at Sundays Well. In 1943, Farrington was appointed Lecturer in Municipal Engineering at UCC, a post he held for many years. The family lived at "Carriglea", Blackrock, county Cork.

On 2 July 1918, Farrington married Evelyn Maud, daughter of William White, a newspaper proprietor from Holywood in county Down.

He was elected MICEI in 1929, became a vice-president 1953, and served as president in 1955. He had previously been president of the Engineers Association in 1945. He was elected AMICE in 1914 and transferred to MICE in 1931.

He died at the Bon Secours hospital in Cork on 24 December 1965, survived by his wife and three daughters.

BUCKLEY, Cornelius John (1899-1968), civil engineer, was born at Main Street, Millstreet, county Cork on 18 June 1899, the son of Michael Buckley, a provisions merchant, and his wife Margaret. Con Buckley was educated at Catholic University School in Lower Leeson Street in Dublin (1910-1917) and entered University College Dublin in 1918 to study civil engineering. On graduating BE from the National University of Ireland in 1922, Buckley worked with the Irish Power Syndicate on a scheme for the hydroelectric development of the river Liffey. After a short period in the Valuation Office in Dublin, he worked on the early stages of the Shannon Scheme, including the culverting of the Blackwater river and Clonlara bridge. Here he met the engineer and artist, Sean Keating, RHA, who was painting elements of the scheme.

In June 1928, he joined the Kenya Public Works Department as an assistant engineer to work near Mombasa on roads and bridges. On his return to Ireland, Buckley became resident engineer on the new Kenmare Bridge and presented a paper on its construction to the ICEI in December, 1933.

He joined the Dublin Port & Docks Board as an assistant engineer in 1935 and took over from Francis Willoughby Bond as Engineer-in-Chief at Dublin Port in 1953. He acted as Chief Fire Officer for Dublin Port during WW2. At the port, he was responsible for the design and erection of a number of large warehouses, including one of four storeys in reinforced concrete, the No.2 graving dock, and the completion of Ocean Pier. Following his retirement in 1965, he worked as a consultant with a Swedish group on the designs for a large harbour in North Africa.

Buckley was married in 1947.

He was elected MICEI in 1931, served on the council of the ICEI, acted for a period as Hon.Secretary, was a vice-president 1953, and served as president in 1956.

Having lived at 2 Victoria Road, Rathgar, Dublin, he died at St Vincent's hospital on 13 August 1968.

BOURKE, Edward Joseph Francis (1906-1996), municipal engineer, was born in Dublin on 4 October 1906, the son of James Bourke, a merchant.

He was educated at O'Connell Schools, Dublin (1918-1924) and entered University College Dublin in 1924, graduating with a BSc and a BE degree in civil engineering from the National University of Ireland in 1927 and ME in 1932. Like many contemporary engineering graduates, he worked first with Siemens Bauunion on the Shannon Scheme before working for a time abroad. He was senior engineer on the aerial survey of Rio de Janeiro (1928-1931) before returning to Europe, where he spent nearly two years as an assistant engineer on the Lochaber power scheme in Scotland.

On returning to Ireland in 1933, he joined Dublin Corporation, initially as a land surveyor, worked with the authority until his retirement in 1971. He began by supervising the construction of housing estates in Crumlin, but probably his greatest contribution was the planning of the water supply and drainage of the city in the years following WW2. In particular, the water supply schemes on the river Liffey at Poulaphuca and Leixslip and the North Dublin and the Greater Dublin drainage schemes allowed the city to expand considerably.

Known to his colleagues as 'Ned', Bourke was appointed in 1950 Dublin City Engineer, Borough Surveyor and Waterworks Engineer, succeeding Norman Chance. Dublin traffic increased greatly during the two decades that Bourke was in charge, and he introduced the system of one-way streets and other traffic management strategies.

On 17 August 1936, Bourke married Mary Agnes, daughter of Gerald Begg, an auctioneer from Fort Ostman, Crumlin in Dublin. They had four daughters and two sons.

Bourke was elected MICEI in 1934, served on council, and was president in 1957.

He died at home in Dublin on 15 July 1996.

O'RIORDAN, Jeremiah Augustine (1904-1985), civil engineer, was born at Myrtle Villa, Northland Road, Derry on 21 November 1904, one of thirteen children of Jeremiah O'Riordan, a senior national schools inspector, and Elizabeth MacCarthy, both from county Cork.

He was educated at Christian Brothers School and Belvedere College, Dublin (1916-1921) and entered University College Dublin in 1921. He first studied mathematics, gaining a BA degree from the National University of Ireland in 1924 and then went on to obtain an MA degree in the following year, after which he studied civil engineering, gaining a BE degree in 1927. On graduating, O'Riordan became a junior assistant engineer on the Shannon hydroelectric project under Professor Frank Rishworth, with responsibility for hydrometric investigations and river gauging.

In 1930, 'Dermot' O'Riordan (as he became known to his colleagues), joined the Hydrometric Survey Office (HSO) attached to the Shannon Power Development. Here he began his long distinguished career in hydrology and hydraulics. When in 1930, the HSO was taken over by the Electricity Supply Board (ESB), he became the Board's Hydrometric Engineer.

By 1934, he was Assistant Chief Civil Engineer and later was appointed Deputy Chief Civil Engineer, a position he held until his retirement in 1969. He lived at 10 Burdett Avenue in Sandycove, county Dublin. During his long career in the ESB, he was particularly involved in the feasibility studies for all the hydropower plants constructed by the ESB and in the design of the hydraulic structures associated with them.

As a young engineer, O'Riordan was responsible for the establishment of a network of river gauging stations on all the rivers in the country that had potential for the production of hydropower, and introduced the first continuous automatic level recorders. This pioneering work, which was later amplified by the Office of Public Works, laid the foundation for the comprehensive records of river flow that are now available.

In the 1960s, O'Riordan became convinced that a pumped-storage scheme should be added to the ESB system. He selected the Lough Nahanagan - Turlough Hill site in county Wicklow and the scheme became operational in 1974. His paper on the project to the ICEI in 1966 won for him the highest and rarely presented Mullins Gold Medal.

On 23 April 1947, O'Riordan married Ellen Mary, daughter of Thomas Cunningham, an egg exporter.

He was elected AMICEI in 1929, transferred to MICEI in 1942, served as Hon Secretary for a number of years, and became president in 1958. He was elected AMICE in 1944.

O'Riordan, who lived in Portobello in Dublin, died of colon cancer at St Vincent's hospital on 4 April 1985.

HARTY, Vernon Dunbavin (1905-1994), civil engineer, was born at "Mountain View", Kenilworth Park, Harold's Cross in Dublin on 17 August 1905, the son of Leonard Dobbin Harty, a commercial clerk from Cork, and Irene Leonora Love.

He was educated at Cork Grammar School, the Manor School, Fermoy, and Cranleigh School in Surrey. He entered Trinity College Dublin in 1924 to study civil engineering under professor David Clark and graduated BA and BAI from the University of Dublin in 1928.

He then spent a short time in the engineer's office of the London North Eastern railway at York before working on the Shannon power scheme from 1928 to 1930. From 1930 to 1931 he was employed by Sir Robert McAlpine in factory construction and from 1931 until 1932 by Sir Cyril Kirkpatrick & Partners on the power house of the Ford Motor Company at Dagenham in Essex. Returning to Ireland, he entered the civil works department of the Electricity Supply Board (ESB) on 30 March 1932 as a member of the hydrometric survey team. He remained in the Board's employment - apart from a short break during WW2 - for the rest of his career, working at Ardnacrusha on the river Shannon and the Pigeon House power station, in Dublin, before becoming resident engineer at the Poulaphuca hydro-electric works, where he remained from 1938 until 1941. From 1941 to 1943, in response to the fuel supply crisis, he was temporarily seconded to the Department of Industry & Commerce to work for the Slieve Ardagh Coalfield Company. In 1946 Harty began a long association with the Erne hydroelectric development scheme, which lasted until its completion in 1956, and he was resident engineer for the construction of the power stations at Cathaleen's Falls and Cliff from April 1946 until September 1950, when he returned to Dublin. Between 1950 and 1959 he was engaged in the construction of coal-fired power stations at Marina in Cork and Ringsend in Dublin, and in the hydroelectric development of the rivers Lee and Clady. In December 1960 he succeeded Joseph MacDonald as the ESB's chief engineer of civil works, and as such was responsible for the new steam power stations at Great Island, Tarbert and Poolbeg.

Following his retirement in 1970 he acted as a consultant to the Central Bank for many years and was appointed arbitrator in numerous engineering contract disputes. He also served on the Board of the Adelaide Hospital and was active as a Friend of St Patrick's Cathedral.

On 4 August 1931, Vernon married Violet Constance Elizabeth, daughter of J. M. Hennelly, a Cork school teacher.

Harty was elected AMICEI in 1934, transferring to MICEI in 1948. He served on the council from 1949 to 1965, and as president 1959/60. He was elected AMICE in 1936.

He died on 9 April 1994 at Newtown Park House nursing home, Blackrock, county Dublin.

COFFEY, Jeremiah Gerard (1908-1988), civil engineer, was born at St Mary's Road, Midleton, county Cork on 27 April 1908, the son of Jeremiah J Coffey, builder and contractor, and Mary Moore.

He was educated at St Colman's College, Fermoy (1921-1925) and at University College Cork, where he studied civil engineering, gaining a BE degree from the National University of Ireland in June 1928.

Following graduation, Coffey acted as clerk-of-works for Carlow county council before being employed on the Ford factory works in Cork. He then carried out additions to Lismore Christian Brothers Monastery for his father's firm (J.J.Coffey & Sons, Midleton) and later a number of artisan dwellings in Midleton. Between 1931 and 1937, he acted as a resident engineer for the South of Ireland Asphalt Co. on a number of waterworks projects, including those at Portarlington, Edenderry, Athboy, Portumna, water supply extensions for Sligo, Cavan, Drogheda and Bundoran, and sewerage extensions in Cavan and Ballinasloe.

On 21 May 1937, Coffey was appointed temporary engineering inspector in the Department of Local Government & Public Health and was assigned to work on water and sewerage schemes and roads. He moved to live at "Santa Cruz", Lower Mounttown Road in Dun Laoghaire. In February 1939, Coffey was seconded to the Department of Industry and Commerce to supervise the construction of the Irish pavilion at the New York World's Fair – the famous 'Shamrock Building' designed by the Irish architect, Michael Scott. Coffey resigned from the Department on 31 March 1940 to become county surveyor for Kilkenny. Much of the work of a county surveyor involved the maintenance and improvement of the road network, and Kilkenny was no exception. In addition Jerry Coffey was responsible for a wide range of engineering work, including water and sewerage and buildings. He retired in 1974.

On 9 October 1940, Coffey married Sheila, daughter of Patrick Glynn, a Dublin motor proprietor.

He was elected AMICEI in 1934, transferred to MICEI in 1938, served on council and became president in 1960.

Coffey died at St Luke's hospital in Kilkenny on 21 December 1988. His brother, Patrick, was professor of engineering at University College Cork until 1967, and a son, Frank, became the first county engineer for South Dublin in 1994.

SIMINGTON, Thomas Aloysius (1905-1994), civil engineer, was born at 92 Phibsborough Road in Dublin on 1 June 1905, the son of John Joseph Simington, an accountant and newspaper manager, and Bridget Mary Mullett.

Thomas was educated at Belvedere College and entered Trinity College Dublin to study civil engineering. He received the degrees of BA and BAI from the University of Dublin in 1927. On graduating, he worked for a time with Great Southern Railways on the design and maintenance of steel structures, before moving to London to work for Tileman & Co. on the design of concrete structures.

In the early 1930s, he was contractor's agent on a variety of bridge, marine and other works. These included a new deepwater jetty at Foynes on the Shannon estuary, extensions at Limerick docks, and Butt Bridge in Dublin – a three-span reinforced concrete cantilevered structure completed in 1932. His next job was the reinforced concrete bow-string girder bridge at Kenmare, designed by Mouchel & Partners, and opened March 1933. He then returned to Dublin to work on the construction of the Cusack Stand at Croke Park (completed 1937, demolished 1993). He worked for Dublin Corporation during WW2 on a variety of projects, and for a time was engineer in charge of relief schemes. In June 1944 he became acting county engineer in Kildare and in September 1945 county engineer for Clare.

After the end of the war, contracting began to pick up and in October 1949, Simington resigned his post in Clare to become joint managing director of the newly formed John Paul Construction Co., the company's first major project being concrete bridge construction associated with the river Lee hydroelectric scheme in county Cork. Simington's subsequent work related mainly to marine and bridge structures, including the reconstruction of the long approach viaduct at Fenit Harbour. He retired in 1972 and lived at "Hibernia", Stillorgan Road, Dublin. He had a life-long interest in bridges and was the co-author, with Peter O'Keeffe, of *Irish Stone Bridges* (Dublin, 1991).

On 25 November 1936, Simington married Alice, daughter of Maurice Power, a Limerick hotel proprietor. They had two daughters, Claire and Ann, and a son John.

He became a MICEI in 1945, served on the council and became president in 1961-62. He was president of The Engineers Association 1959-60.

'Tommy' Simington died at St Vincent's hospital in Dublin on 17 June 1994 and is buried in Deansgrange cemetery.

O'CONNOR, Thomas Joseph (1902-1984), civil engineer, was born at Mitchell Street, Clonmel on 19 March 1902, the son of Edward O'Connor, a butcher, and Elizabeth (Lizzie) Scott. Thomas was educated at High School, Clonmel (1907-1919) and at University College Dublin, where he studied civil engineering, gaining a BE degree from the National University of Ireland in 1923, having been awarded a Pierce Malone scholarship. He also obtained a BSc degree from the Royal College of Science in Ireland.

In 1925, O'Connor accepted a two-year position as Assistant to Professor Hummel in the civil engineering department of Queen's University in Belfast, where he taught surveying and sanitary engineering, including hydraulics. He later worked for Siemens Bauunion on the construction of the Shannon scheme at Ardnacrusha. From 1929 until 1933 he worked in Argentina on behalf of the Greenly Commission (a US development agency) assessing routes for new railways.

Returning to Ireland, he worked for a time in Derry with Michael O'Doherty and subsequently with the consulting engineer, Patrick Hartnett McCarthy in Dublin on a variety of water supply and sewerage projects, including Dingle water supply and Mullingar sewerage schemes. Surveys and preliminary designs for water supply schemes for Arklow and Rathdrum followed. He worked on the preparation of plans for the Liffey hydroelectric scheme at Poulaphuca and assisted Edward M.Murphy with the plans for water supply schemes for Balbriggan and Skerries.

In 1937, O'Connor set up his own consulting engineering practice in Commercial Buildings, Dame Street, Dublin, specialising in civil and structural engineering, with particular emphasis on water services. The name of the firm was subsequently changed to T J O'Connor & Associates.

He was elected AMICEI in 1925, later transferring to MICEI. He served as ICEI president 1962-63 and in 1961 as president of the Association of Consulting Engineers of Ireland.

O'Connor, who had been a resident of Killincarrig House in Greystones, died of prostate cancer, at St Vincent's hospital in Dublin on 17 August 1984, predeceased by his wife.

LANE, John (1906-1991), civil engineer, was born at Berrings, Inniscarra, county Cork on 17 December 1906, the eldest son of Eugene Lane, a farmer, and Margaret Kellegher.

He was educated at Berrings North School and the CBS College in Cork. He entered University College Cork in 1924 to study civil engineering under Professor Harry Walsh, graduating BE from the National University of Ireland in 1927.

John was first employed as clerk-of-works on road construction schemes for Cork county council under Richard Francis O'Connor. He then worked as an assistant engineer under Frank Rishworth on the Shannon Scheme.

In October 1930 he became an assistant engineer with the Electricity Supply Board, carrying out surveys in Lough Allen. From June 1931, Lane worked for the South of Ireland Asphalt Co., where he was responsible for roads, sewerage and water supply contracts. Between 1944 and 1947, he worked for the Department of Local Government.

On 16 December, 1947, he was appointed county engineer for Monaghan, where he served for fifteen months before, in March 1949, being appointed county engineer for Laois, where he was to serve for the next 24 years. He had a particular concern for the health of the network of bridges in the county. During his time with both county councils, Lane continued to live at his Dublin home in Orwell Gardens in Churchtown. While in Monaghan, he returned home most weekends, but commuted daily by train to Portlaoise when working for Laois county council. He retired in 1971.

On 24 October 1936, John Lane married Eileen Ita, daughter of John and Eileen O'Connor of Cloghroe, county Cork. They had four children, Owen, Ciaran, Patricia and Jim, the last named also becoming a civil engineer.

John Lane was elected AMICEI in 1938, transferred to MICEI in 1947, was a member of council, and served as president in 1963. He died at Dublin on 14 April 1991, predeceased by his wife, who died 1 April 1991. They are buried in Bohernabreena cemetery in county Dublin.

KELLY, Thomas (1908-1984), civil engineer, was born at Aylwardstown House, Kilmackevogue, county Kilkenny on 8 November 1908, the son of Nicholas Joseph Kelly, a farmer, and Catherine Dooley. He was educated at Patrician College, Mountrath and entered University College Dublin in 1926. He received the degrees of BE and BSc from the National University of Ireland in 1929.

He was a student engineer with Siemens Bauunion on the Shannon Scheme in 1928 and in 1929-30 undertook a post-graduate course in geology. He was employed in 1930-31 by D.S.Doyle, consulting engineer, as clerk-of-works on the construction of the Clonmel mental hospital. In 1931-32, he was engineering assistant to William J Doherty of Derry, working mainly on water and sewerage schemes, and later as clerk-of-works on another hospital project.

In July 1932, Kelly became a temporary assistant county surveyor with Kilkenny county council, and permanent assistant in September 1935. He served also as engineer to the Kilkenny Board of Health. In 1943, he was placed first in the interviews for the new post of County Engineer in Wexford and took up the post in June of the following year. In November 1945 he transferred to the county Kildare authority where he was responsible for commencing the major realignment and other improvement works on the Dublin – Naas road, which led to the opening in June 1963 of the first three-mile section of dual carriageway.

In 1959, following the resolution of a dispute over basic salaries, Kelly was appointed to the vacant post of County Engineer for Dublin. Whilst developing the council's own road improvement programme, he put forward in 1961 a well-argued case for the establishment of a central main road authority. He retired in 1962 due to ill health.

He became a student member of the ICEI in 1927, was elected AMICEI in 1931, transferred to MICEI in 1941, served on council, and was president in 1964-5. He was also president of the Engineers Association in 1960 and in the same year, district chairman of the Institution of Municipal Engineers.

Having lived at 30 Belgrove Road, Rathmines, Tom Kelly died at Our Lady's Hospice in Harold's Cross on 12 May 1984 of prostate cancer.

CROSS, Richard Ernest (1901-1980), civil engineer, was born at 18 Sullivan's Quay in Cork on 12 January 1901, the youngest son of Thomas Joseph Cross, carriage builder, and Margaret Flynn from county Waterford.

He was educated at CBS North Monastery, Cork (1910-18) and entered University College Cork (UCC) in 1919, where he first studied mechanical and electrical engineering, gaining a BE degree from the National University of Ireland in 1922. He then worked for some years in the power generation and distribution department of the Ford factory in Cork before returning to UCC to study civil engineering, graduating for a second time in 1925.

Cross then joined the Office of Public Works (OPW) as assistant to Thomas Batchen, where he was to serve for over forty years until his retirement in 1966. During this period the OPW undertook a comprehensive national programme of arterial drainage schemes and Cross was identified closely with these developments, such as the Barrow catchment scheme, where he was assistant to John Chaloner Smith (1933-36) and the Erne scheme, where he acted as resident engineer (1936-38). Cross was in charge of the development of hydrometric surveys of Irish rivers and from 1943 to 1948 was involved in the design of drainage works on the Brosna, Glyde/Dee and Feale catchments. From 1948 to 1954, he was Engineer-in-Charge of work on all catchment schemes carried out by the OPW by direct labour. These works included the construction of river and sea embankments, new bridges, weirs, pumping stations, and the underpinning of hundreds of existing bridges.

In 1954, Cross was appointed Assistant Chief Engineer at the OPW with responsibility for the survey, design and execution by direct labour of all works of land drainage and reclamation. In 1960 he was appointed Chief Engineer of the OPW, thus to arterial drainage was added responsibility for the development and maintenance of the state harbours, fishing ports, and coastal protection.

On 22 November 1932, Cross married Catherine Mary O'Sullivan, daughter of Patrick O'Sullivan of Rathpeacon House, Blarney, county Cork.

He was elected AMICEI in 1933 and transferred to MICEI in 1962, was a member of council, was elected a vice-president in 1964 and served as president in 1965-6. He was elected AMICE in 1933. His interest in archaeology, as a result of discoveries during drainage works, led to him being awarded honorary fellowship of the Royal Society of Antiquaries. As somewhat of a bibliophile, he was a member of the library committee of the Royal Dublin Society. He was also a founder member of An Taisce (the National Trust for Ireland) and a member of its council. He was a founder of the Institute of Public Administration and a member of its council for many years.

'Ernie' Cross, whose residence had been at 188 Kimmage Road West in Dublin, died at Jervis Street private hospital on 6 May 1980.

140

HERLIHY, Daniel (1909-1982), civil engineer, was born in county Cork in 1908, the son of Jeremiah Joseph Herlihy, a civil servant. Daniel was educated at North Monastery CBS, Cork (1925-27) and at University College Cork, graduating with a BE degree in civil engineering from the National University of Ireland in 1931. On graduating, he obtained a scholarship tenable in the engineering department of the Cork Harbour Commissioners. Soon after his period with the commissioners, Herlihy joined the local government service and was employed first by Tipperary (NR) council on a number of projects. In 1932/3 he was engaged on bridge repair and reconstruction. During 1933-34, he was resident engineer (housing) with Thurles RDC. This was followed by the design and supervision of a sewage scheme for the Board of Health and on the construction of concrete roads in the county. In 1934 he was appointed Assistant County Engineer for Tipperary (NR) and Town Surveyor of Templemore.

In 1937, Herlihy was appointed to an engineering inspector position in the department of Local Government and was promoted successively to the posts of Senior Engineering Inspector (1941) and Assistant Chief Engineering Advisor (1947), dealing mostly with the national roads system. During his sixteen years in local government, Herlihy was closely associated with many important developments and presided over numerous public inquiries. He served on many inter-departmental committees, in particular he was noted for his work on the committee dealing with employment schemes. With his wide experience of local government, he later was called upon to lecture on courses leading to a Diploma in Public Administration.

In 1950 Herlihy was seconded by the Department of Agriculture for four months to investigate a proposal that Córas Iompair Éireann (CIE) become the national operator of a new government scheme for transporting and spreading ground limestone. In the same year, he was appointed chief engineer of CIE and deputy general manager, and the ground limestone scheme became a significant company activity for many years. Later, in 1970, Herlihy became general manager. Possibly his major achievement during this period was the impetus given to planning the use of new technology to adapt existing services, resulting in RailPlan80, announced shortly following his retirement on health grounds in August 1972.

On 25 April, 1935, Dan Herlihy married Mary Brigid, daughter of John Hanly, a farmer from near Nenagh in county Tipperary. They lived at Rockfort, Dalkey in the 1960s but later moved to Foxrock. They had sons, Diarmuid and John.

Herlihy was elected AMICEI in 1934, transferred to MICEI in 1940, served on council, was vice-president 1964 and served as president 1966-67. He was also an associate member of the Town Planning Institute, and a member of the Institute of Transport, of which he was chairman of the Irish section in 1968.

Dan Herlihy died at 43 Hainault Road, Foxrock, county Dublin on 24 November, 1982.

RAFTERY, Patrick Joseph (1920-1986), civil engineer, was born at Abbey Street, Roscommon on 25 August 1920, the eldest son of Patrick Joseph Raftery and Margaret Halligan. Shortly after his birth, the family moved to Dublin when his father became an engineering inspector in the Department of Local Government.

Patrick was educated at the Christian Brothers School in Synge Street (1932-39) before entering Trinity College Dublin to study civil engineering. He gained the degrees of BA and BAI from the University of Dublin in 1942 and was later awarded the higher degrees of MA and MAI.

He began his career as an assistant to Professor John Purser in the TCD engineering school, where he was demonstrator in geology and in surveying. Raftery then held various posts in the local authority service: with Dublin Corporation, Dublin county council, and Meath county council, where he was Chief Assistant County Engineer, before moving in 1962 to a similar post with Dublin county council. He worked on the Howth outfall tunnel, part of the North Dublin drainage scheme, and other elements of the drainage of Dublin, as well as roads and other aspects of the engineering of the city. He lived at 64 Upper Leeson Street.

Following in the footsteps of his father, he became an engineering inspector in the Department of Local Government in 1965. Previously, in 1961, he had been called to the Bar at King's Inns, Dublin. With this rare combination of experience and qualifications, he was called upon by the Department of Energy to conduct the hearings for the acquisition of the right-of-way for the natural gas pipeline from Cork to Dublin. His legal and engineering expertise was also put to good use in the drafting of the building regulations and guidelines. He retired from what became the Department of the Environment in 1985.

He was elected AMICEI in 1945 and was later transferred to MICEI. He served on council, was a vice-president and served as president 1967-68. He was elected AMICE in 1948. He was also an AMIMunE.

Outside of the engineering profession, he had an enormous involvement in cultural and educational activities, and for many years served as Secretary of the St Vincent de Paul Council for Ireland.

He died, unmarried, at Baggot Street hospital on 21 December 1986, having resided at 14 Palmerston Road in Dublin.

DOOGE, James Clement Ignatius (1922-2010), civil engineer, was born at 78 Grange Road West, Birkenhead on 30 July 1922, the son of Denis Patrick Dooge, an Irish marine engineer in the British merchant navy, and Veronica Carroll.

He was educated, first in Liverpool, and then at CBS, Eblana Avenue, Dun Laoghaire (1932-39). He entered University College Dublin (UCD) in 1939 to study civil engineering and graduated BE from the National University of Ireland (NUI) in 1942.

His was first employed as clerk-of-works on the wastewater treatment plant for Peamount Sanatorium before moving to the Office of Public Works in March 1943 to work on planning and design of arterial works under Dr J J Kelly. In 1946, Dooge joined the Electricity Supply Board (ESB) where for the next ten years he worked on studies and project planning for hydroelectric development. It was while with the ESB that he developed his deep interest in hydrology.

He was awarded an ME degree in civil engineering from the NUI in 1952, his thesis being "unsteady flow in open channels". He then took two years leave of absence to become a research associate at Iowa State University, where he obtained an MSc in Fluid Mechanics and Hydraulics. From March 1956 until 1958, he was senior design engineer with the ESB prior to becoming Professor of Civil Engineering at University College Cork, where he was to remain for twelve fruitful years. During this time he became a specialist consultant in the area of engineering hydrology, water and drainage surveys and design, and tidal flooding. He led many advances in the application of linear systems theory to hydrology in general and rainfall run-off modeling processes in particular.

In 1970, Jim Dooge was appointed Professor of Civil Engineering at UCD, remaining in the post until his retirement in 1984. He was research consultant at the Centre for Water Resources Research at UCD and was also for a time a visiting professor at the Department of Engineering Hydrology at University College Galway. He lived at 25 Villarea Park in Glenageary, county Dublin.

During his career, he became interested in politics and was elected in 1948 to Dublin County Council on a Fine Gael ticket and was twice chairman of the council. In 1961 he was elected to the Irish Senate, having been nominated by The Engineers Association. In 1973 he became chairman of the Senate and a member of the Council of State. In 1981 he was appointed by the then Taoiseach, Garrett Fitzgerald, as Minister for Foreign Affairs, but his tenure of office was brief due a change of government. In 1983, he was a member of the New Ireland Forum and the following year, chairman of the ad-hoc committee of representatives of EEC heads of government on institutional affairs and the European Union whose report formed the basis of the Single European Act. He retired from public life in 1987.

Professor Dooge became involved with a myriad of other international organisations, committees and working groups sponsored by the European Union and the United Nations in the areas of world climate, water resources and geo-sciences. He served as chairman of the Scientific Advisory Committee of the World Climate Impact Studies Programme 1980-90; as a member of the International Decade for natural Disasters Reduction advisory board on behalf of the Secretary-General of the UN; served as president of the International Commission on Water resources 1971-75; as president of the International Association for Hydrological Sciences 1975-79; and as president of the International Council of Scientific Unions. In a tribute paid to him by UNESCO on the occasion of his death, he was described as a 'towering figure and pioneer in hydrology'.

On 25 November 1946, Dooge married Marie Veronica (Roni), daughter of Joseph O'Doherty, a Dublin accountant. They had two sons and three daughters.

Professor Dooge was elected MICEI in 1947, FICEI in 1957, served on council from 1958 to 1973, becoming president in 1968-69, the session during which the merger took place between Cumann na nInealtóirí (The Engineers Association) and the ICEI. He played a leading part in the merger negotiations with Jock Harbison and others and steered the required legislation through the Oireachtas. For his presented papers to the Institution, he was awarded a Kettle Premium and Plaque, and on two occasions, a Mullins Silver Medal. He was later awarded honorary fellowship of Engineers Ireland and was a founder fellow of the Irish Academy of Engineering. He was elected AMICE in 1959, and transferred to MICE in 1964. He was also a EurIng and a fellow of the ASCE.

Dr Dooge was Secretary of the Royal Irish Academy (RIA) 1978-1981 and was the first ICEI president to be elected president of the RIA (1987-90), from which body he received a Gold Medal. In 2000, he was elected a fellow of the Royal Academy of Engineering and was awarded their Prince Philip Medal in 2005 as 'an outstanding figure in the field of hydrology'. Professor Dooge published over 100 papers as sole author and an equal number as joint author. He was awarded a number of honorary doctorates: from the University of Wageningen (Agriculture Sciences, 1978), from the University of Lund (Technology, 1980), from the University of Birmingham (Science, 1988), and a DSc from the University of Dublin in 1988.

He died at his residence in Monkstown, county Dublin on 20 August 2010, predeceased by his wife, and is buried at Deansgrange cemetery.

Reference Sources

Books:

Beesley, G (in preparation) *A biographical dictionary of Irish railway engineers*

Boase, F (1897 and 1901) *Modern English Biography,…*Truro, Netherton and Worth

Boylan, H (1978) *A dictionary of Irish biography.* Dublin, Gill and Macmillan

Burtchaell, G.D and Sadleir, T.U (1935) *Alumni Dublinensis*, New Revised Edition, Dublin, Alex. Thom & Co.,Ltd. [also volumes entitled *A catalogue of graduates of the University of Dublin*, V2 1896, V3 1906, V4 1917]

Corcoran, M (2005) *Our good health: A history of Dublin's water and drainage*. Dublin, Dublin City Council

Cosgrave, E.M (1908) *Dublin and County Dublin in the Twentieth Century*. Brighton, W.T.Pike & Co.

Cox, R.C (1993*) Engineering at Trinity, incorporating the 7th Ed. of the Record of the School of Engineering,* Dublin, TCD School of Engineering, Trinity College Dublin

Cox, R.C (2009) *Civil Engineering at Trinity*. Dublin, Department of Civil, Structural & Environmental Engineering, Trinity College Dublin

Cox, R.C (Ed.) (2006) *Engineering Ireland*, Cork, Collins Press

Cox, R.C and M.H.Gould (1998) *Civil Engineering Heritage : Ireland*. London, Thomas Telford Publications

Cox, R and P.Donald (2013) *Ireland's Civil Engineering Heritage*, Cork, Collins Press

Crone, J.S (1928) *A concise dictionary of Irish biography*, Dublin, Talbot Press

Cross-Rudkin, P.S.M and Chrimes, M.M (2008) *Biographical dictionary of civil engineers in Great Britain and Ireland: Volume 2: 1830-1890*, London, Thomas Telford Publishing

Duffy, P 'Engineering' in Foley, T (1999) *From Queen's College to National University*, Dublin, Four Courts Press

Gilligan, H.A (1988) *A history of the port of Dublin*. Dublin Gill and Macmillan

Griffith, A.R.G (1987) *The Irish Board of Works 1831-1878*, New York/London, Garland

Hawkins, R (2007) *Dictionary of Irish Biography*, Oxford, Royal Irish Academy

Marshall, J.A (1978) *A biographical dictionary of railway engineers*. Newton Abbot, David and Charles

O'Donoghue, B (2007) *The Irish County Surveyors 1834 – 1944: A biographical dictionary,* Dublin, Four Courts Press

O'Riain, M (1995) *On the move: Córas Iompair Éireann 1945-1995*, Dublin, Gill and Macmillan

Pike, W.T (Ed.) (1910) *British Engineers and Allied Professions in the Twentieth Century*. Brighton, W.T.Pike & Co.

Robins, J (1993) *Custom House People*, Dublin, Institute of Public Administration

Skempton, A.W et al (2002) *A biographical dictionary of civil engineers in Great Britain and Ireland; Volume 1: 1500-1830*, London, Thomas Telford Publishing

Steer, F.W.et al (1997) *Dictionary of Land Surveyors and Local Map-Makers of Great Britain and Ireland 1530-1850*. London, British Library

Online resources:

Rowan, Ann-Martha (2007) *Dictionary of Irish Architects*, Dublin, Irish Architectural Archive [accessed online at www.dia.ie]

Oxford Dictionary of National Biography, Oxford University Press, 2004 [accessed online at www.oxforddnb.com, via Trinity College Dublin Library]

Newspaper archives accessed online via TCD Library included:

The Irish Times

The Irish Independent

The Freeman's Journal

The Times

Journals:

Empire Survey Review

Engineering

Engineers Journal

ESB Staff Journal

Illustrated London News

Institution of Mechanical Engineers Proceedings

Irish Engineering Journal

Journal of the Institution of Civil Engineers

Journal of the Institution of Electrical Engineers

Minutes of the Proceedings of the Institution of Civil Engineers

Proceedings of the Royal Society

Professional Papers of the Corps of Royal Engineers
The Dublin Builder
The Engineer
The North Munster Antiquarian Journal
Transactions of the Institution of Civil Engineers of Ireland

Genealogical Sources:

Indexes to UK and Irish birth, marriage, and death records and UK Census 1841, 1851, 1861, 1871, 1881, 1891, 1901, 1911 accessed through www.ancestry.co.uk. Certificates obtained from General Registry Office, Southport and the General Register Office, Dublin.
Irish Census 1901, 1911 [accessed online at www.census.nationalarchives.ie]

Other sources:

Dublin and Irish street directories
ICEI membership application forms
ICE membership application forms
IMechE membership application forms
IEE membership application forms